FUNGI AND THEIR UTILIZATIONS

FUNGI AND THEIR UTILIZATIONS

Rabia Iftikhar

Delve
PUBLISHING

www.delvepublishing.com

Fungi and their Utilizations

Rabia Iftikhar

Delve Publishing
2010 Winston Park Drive,
2nd Floor
Oakville, ON L6H 5R7
Canada
www.delvepublishing.com
Tel: 001-289-291-7705
001-905-616-2116
Fax: 001-289-291-7601
Email: orders@arclereducation.com

ISBN: 978-1-77361-057-3(Hardcover)

ABOUT THE AUTHOR

Ms. Rabia Iftikhar has an M.Phil Degree from Lahore University of Management Sciences (LUMS) in 2016. Currently she is working as an IGCSE Biology Teacher at Lahore Grammar School (Johar Town Boys Campus), Lahore, Pakistan. Previously, she worked as a teacher at Bangladesh International School Riyadh, Saudi Arabia.

TABLE OF CONTENTS

LIST OF FIGURES

LIST OF TABLES

LIST OF ABBREVIATIONS

2-DE	two-dimensional electrophoresis
2D-DIGE	2D-difference gel electrophoresis
2D-PAGE	two-dimensional polyacrylamide gel electrophoresis
6-APA	6-amino penicillanic acid
7-ACA	7-aminocephalosporonic acid
ADH1	alcohol dehydrogenase 1
BCA	biocontrol agents
CGD	candida genome database
DTT	dithiothreitol
FGC	fungal genomes central
FT-ICR	Fourier-transform ion cyclotron resonance (FT-ICR)
GAL1	galactokinase gene
GAPDH	glyceraldehyde-3-phosphate dehydrogenase
HCCA	α-cyano-4-hydroxycinnamic acid
IEF	isoelectric focusing
LSD	lysergic acid diethylamide
MALDI-ToF ICMS	matrix-assisted laser desorption ionization–time-of-flight intact cell mass spectrometry
MALDI-ToF MS	matrix-assisted laser desorption ionization–time-of-flight mass spectrometry
MS	mass spectrometry
MSA	multiple sequence alignment
NAG	N-acetylglucosamine
NAM	N-acetylmuramic acid
nanoESI	nanoelectrospray ionization
NCBI	The National Centre for Biotechnology Information

NGS	next-generation sequencing
ORE	estrogen receptor element
PAGE	polyacrylamide gel electrophoresis
PBP	penicillin-binding proteins
PGK	phosphoglycerate kinase
PTMs	post-translational modifications
PVDF	polyvinylidene fluoride
REMI	restriction enzyme-mediated insertions
ROS	reactive oxygen species
RT-PCR	reverse transcriptase- polymerase chain reaction
SCP	Single-cell protein
SDS-PAGE	sodium dodecyl sulfate-polyacrylamide gel electrophoresis
SGD	saccharomyces genome database
TCA	trichloroacetic acid
YEP	yeast episomal plasmid

PREFACE

The kingdom of fungi consists of several diverse groups of organisms. They are abundant, numerous, and carry out many functions linked to other organisms as well as independently. The species of fungi vary widely in sizes from those that are few micrometers in size to those that are much larger and are made up of fruiting bodies that have sizes in range from few centimeters to even meters. In some extreme cases, fungal species can develop and colonize into areas as large as hectares. Not only in structural form, this kingdom also varies widely in their functions. They can literally inhabit any ecological niche including those that harbor slow-growing endolithic communities present in the Polar regions and those that are pathogenic to plants and animals and are able to degrade organic material within the tropical regions and are highly specific. Due to this wide diversity in functions exhibited by the fungal kingdom, they are widely employed in cell biology. Recent advances in techniques in the field of molecular biology; amplification of DNA, systems for gene cloning and expression, analysis of the genome and proteome, together have enabled detection of several novel fungal compounds. In additional these available tools have also allowed investigation, characterization, and exploitation of novel and old known fungal activities. For many industrial processes that are based on biotechnology and for development of various compounds, fungal species have played a key role. Fungi have also been targeted for numerous biotechnological applications inclusive of those that serve to develop and produce significant pharmaceutical and industrial metabolites where they provided systems for heterologous and homologous over-expression of genes.

There is a wide variety of literature available which covers numerous aspects of general fungal study (mycology). More specifically, these literature data belongs to areas such as biochemistry, molecular biology, ecology, systematics, and pathology. Currently, there is not much literature available related to the basic principles of applied mycology and fungal biotechnology. Their diverse implications are broken down and published within the literature in the form of some specialized journals. I wrote this book with the intention to compile this diverse field in one book "Fungi and their Utilizations." This book broadly covers the basic principles, methods, and applications of mycology, treating it as an integral subject in various disciplinary subjects. This book presents and collates the important aspects of applied mycology and serves as a tool for students, teachers, and researchers. The developments that have occurred in the fields of biological sciences and industries have accelerated, especially with respect to the physiology, ecology and biochemical processes in fungi. Such information is a key to biotechnology and can be manipulated for usage. In this book, the current developments have been reviewed and the recent advances have been highlighted especially in the areas of molecular biotechnology which have application in medical biotechnology, pharmaceutical industry, agricultural biotechnology, commercial production technology and in industry.

This book is intended to serve as a source of information related to an extensive introduction of various aspects of fungal biotechnology. It has chapters that cover updated information about the molecular and biochemical techniques that are developed to facilitate, fungal genomic, post-genomic and proteomic analyses. For example, it includes details about the various bioinformatics tools that are applicable in fungal studies, genetic screening, mutant studies, DNA microarrays and a lot more. It is targeted to present fungal kingdom as the model organisms for use in latest developed techniques. Moreover, it also includes how these developments and advancements have been used to exploit the wide range of metabolites that are synthesized by the fungi. One of the chapters in this book is completely dedicated to the diverse classes of enzymes that are produced by several fungal species and are utilized in several industries including foods, feeds, detergents, beverages, baking etc. Some chapters are included that have direct or indirect implications in biotechnology.

The huge variety of uses and properties of fungi offers innumerable ways by which humans can exploits these fungal characteristics for their own benefits. Therefore, techniques of fungal characterization, preservations, recent advances in fungal biotechnology, access to in silico database and genomic information are all worth sharing. All these challenging aspects related to fungal studies are also covered in this book. The potential of biofungicides, production of edible fungi, fermented foods, high-value products like mycoprotein, mycoherbicides and bioherbicides are also briefly discussed. The aim of this book is to provide a detailed source of information for fungi, their biology and their utilization in various key fields including agriculture, food, etc.

Conclusively, this book is intended to provide useful resources for students, teachers, and researchers as contains a comprehensive detail about latest applications of fungi. I expect it to be beneficial to them in understanding and applying novel technologies in the field of mycology.

1
CHAPTER

AN INTRODUCTION TO FUNGAL PHYSIOLOGY

CONTENTS

1.1. INTRODUCTION

Physiology of a cell includes its nutrition, metabolism, growth, reproduction, and death. Physiology of fungi also involves its interactions and responses to both biotic and abiotic factors of the environment such as stress. Fungal cell physiology has a significant impact on various industrial processes, the surrounding environment, and human health. A major ecological role is played by fungi in biogeochemical cycling of nutrients. These organisms act as decomposers, i.e., they break down the dead and decaying matter. Fungi also play significant mutual symbiotic, pathogenic and saprophytic roles. They mobilize various nutrients and alter the physicochemical environment within their surroundings. Moreover, fungal metabolism aids in the neutralization of organic pollutants and in bioremediation of heavy metals found in the environment. A number of economically significant industrial commodities are produced by utilization of yeast and fungal metabolism. Some examples include whole foods, food additives, fermented beverages, antibiotics, probiotics, pigments, pharmaceuticals, biofuels, enzymes, vitamins, organic and fatty acids and sterols. Fungi also have the ability to influence human health, both positively and negatively. While some act as pathogens to humans, others function as antimicrobial and chemotherapeutic agents. In the recent years, advances in biotechnology have resulted in techniques that utilize some species of yeast as hosts for the expression of human therapeutic proteins following recombinant DNA technology. Apart from their direct industrial role, fungi and yeast, particularly the yeast *Saccharomyces cerevisiae*, have also served as model organisms for research in the fields of biological and biomedical science. This exploitation has increased remarkably after obtaining the genome sequencing data of multiple fungal species. Both the genomic and proteomic data from fungal species has helped shed light on the insights of human genetics and inherited disorders. For the elucidation of the presently unknown genes of fungi and for identification of their roles, the study of cell physiology is an essential component. Conclusively, fungi are vital organisms for human society, health, and well-being. The understanding, control, and exploitation of fungi are, therefore, fundamentally dependent on the study of their physiology. In this chapter, we will be discussing the basic characteristics of fungal cell physiology, with a primary focus on nutrition, growth, and metabolism in unicellular yeasts and filamentous fungi.

1.2. MORPHOLOGY OF YEASTS AND FUNGI

There are many variations that are observed in the morphologies of yeast and fungi, however, most of the fungi are filamentous. They are often observed as unicellular yeasts, while some primitive species of fungi, like chytridomycetes. These fungi grow as either rounded separate cells or them as dichotomous branched chains of cells consisting of root-like rhizoids that aid in attaching the cells to nutrient resources. In this chapter, we will discuss the two most prevalent morphological forms, filamentous fungi, and unicellular yeasts.

1.2.1. Filamentous Fungi

There are a wide variety of morphologies that are observed among macrofungi and microfungi. A diverse range of mushrooms and toadstools and macrofungi that differ in the morphology of their sexual fruiting bodies are often seen. Microfungi, also called as molds, are also grown in a range of morphologies on decomposing foods and detritus. Some of them are common plant pathogens including the colored rusts, smuts and mildews. The long branched thread-like structures that are visible on molds are termed as hyphae. These aggregates are organized to aid reproduction and dissemination through the formation of spores. A substratum, organized in the form of a network called as mycelium, supports the growth of hyphae. These outgrowths spread and branch apically within this network forming filaments that helps in exploiting and translocating nutrients that are available in the surroundings. (Figure 1.1) Such fungi that grow with the help of formation of hyphae are said to be eucarpic. Depending on the type of species of fungi or conditions of growth, the growing hyphae vary in diameters in the range of around 1 to 30 μm. In a laboratory, the filamentous fungi may proliferate on both liquid or solid culture media. When grown on solid agar media, the colonies of filamentous fungi are seen grown radially and expanding to form clearly visible mycelium. These structures initiate to synthesize from an inoculum, resulting in formation of colonies that are spherical and three-dimensional on, within, or above the substrate.

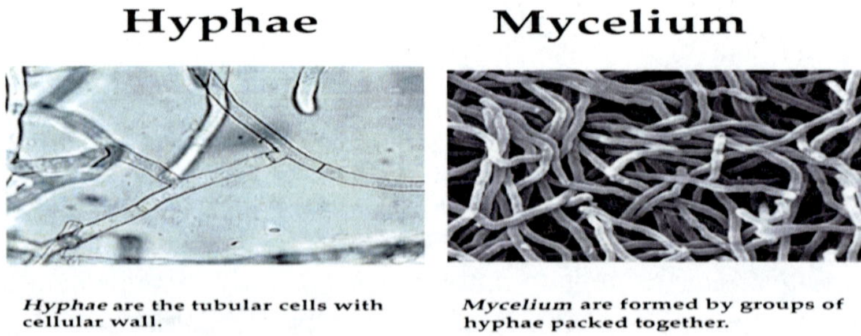

Figure 1.1: a) Hyphae: tubular cells with cell wall, b) Mycelium: Groups of hyphae packed together (Source: *http://www.davidmoore.org.uk/Sec01_03.html*).

Theoretically, the hyphae of fungi may be allowed to apically grow indefinitely if supplied endlessly with sufficient nutrients under ideal environmental conditions. The eucapric fungi are known to be indeterminate organisms, both spatially and temporally. Additionally, contrasting to plants, animals, and other microorganisms, these organisms do not grow up to a limit of predetermined size or age. The structure of mycelium is not homogeneous, it shows considerable variations during their development. Depending on the environmental conditions in which these fungi are kept, their mycelium may adopt interconnections which may grow, branch, fuse, age, die, sporulate either simultaneously or at different times. They may, therefore, display variations in both physiological and biochemical properties. If inoculated on a homogeneous nutrient media, pigmented colonies may propagate, or the colonies may display various morphological sectors, yield aerial structures, develop as either fast-effuse forms or as slow-dense forms. They may even exhibit rhythmic growth. Some higher fungi like basidiomycetes, when incubated on a heterogeneous nutrient media, produce reproductive structures and substrate mycelium, along with long thread-like structures called rhizomorphs or cords. The role of these developed organs is to explore, connector transport water and nutrients across different areas of closely located resources. These linear entities may often mature to develop internal vessel hyphae within them. These provide channels of wide diameters that run along the rhizomorphs. Hyphae that are located peripherally are usually more tightly packed and produce melanin which aids in insulation.

Filamentous fungi and yeasts basically develop as a result of occupying altered habitats and diverse growth polarities. These variations simply result in divergent growth styles of these microorganisms. Several species, termed as dimorphic fungi, have the ability to develop as hyphal or unicellular yeast forms depending on the environmental conditions. For instance, few significant plant and animal pathogenic fungi develop as yeast forms that are able to survive within body fluids. To help them in invading tissues, these fungi also have the ability to grow hyphae or pseudohyphae.

1.2.2. Yeasts

Yeasts are a type of fungi that are unicellular and have diameters ranging from 1–10 μm and in lengths from 2–3 μm to 20–50 μm. These fungi reproduce asexually through the process of budding or fission. *Saccharomyces cerevisiae* is a type of yeast that is ellipsoidal in shape and has a large diameter in the range of 5 to 10 μm and the small diameter of 1 to 7 μm. These yeasts are often referred to as brewer's or baker's yeast. The colonies of yeasts that are grown on agar medium have variations in their colors, textures, and geometries. Pigmentation is often observed in yeasts colonies that are grown on agar surfaces. Commonly observed colors are cream (e.g., *S. Cerevisiae*); white (e.g., *Geotrichum candidum*); black (e.g.*Aureobasidium pullulans*); pink (e.g., *Phaffia rhodozyma*); red (e.g., *Rhodotorularubra*); orange (e.g., *Rhodosporidium* spp.); and yellow (e.g., *Cryptococcus laurentii*). Sometimes, the pigments produced by yeasts are utilized in biotechnological fields. For example, the feed supplements that are used for salmon farming contain astaxanthin produced by *P. rhodozyma.* A huge diversity is observed in yeast shapes which have been summarized in Table 1.1.

Table 1.1: Diversity in Shapes of Yeast Cells

Cell shape	Description	Examples of yeast genera
Ellipsoid	Ovoid shaped cells	*Saccharomyces*
Cylindrical	Elongated cells with hemispherical ends	*Schizosaccharomyces*
Apiculate	Lemon shaped	*Hanseniaspora, Saccharomy-codes*
Ogival	Elongated cell rounded at one end and pointed at other	*Dekkera, Brettanomyces*
Flask-shaped	Cells dividing by bud-fission	*Pityrosporum*

Miscellaneous Shapes	Triangular Curved Spherical Stalked	*Trigonopsis* *Cryptococcus* (e.g., *Cryptococcus cereus*) *Debaryomyces* *Sterigmatomyces*
Pseudohyphal	Chains of budding yeast cells which have elongated without detachment	*Candida* (e.g., *Candida albicans*)
Hyphal	Branched or unbranched filamentous cells which form from germ tubes. Septa may be laid down by the continuously extending hyphal tip. Hyphae may give rise to blastospores	*Candida albicans*
Dimorphic	Yeasts that grow vegetatively in either yeast or filamentous (hyphal or pseudohyphal) form	*Candida albicans* *Saccharomycopsis fibuligera* *Kluyveromyces marxianus* *Malassezia furfur* *Yarrowia lipolytica* *Histoplasma capsulatum*

Source: Walker, G. M., & White, N. A. (2005). Introduction to Fungal Physiology. Fungi. John Wiley & Sons, Ltd.

1.3. THE STRUCTURAL COMPONENTS AND FUNCTIONS OF FUNGI

1.3.1. Cell Surface of the Fungi

The peripheral structure of both the yeast and fungi is termed as the cell envelope. It encloses the cytoplasm and is composed of a plasma membrane, periplasm, cell wall and any additional extracellular structures like fimbriae and capsules. An exoskeleton that protects the protoplast of the fungi from the outside environment is the cell wall. It also functions to direct growth, strengthens the cell, provide shape and other interactive properties to the fungal cell. The process of apical growth is closely related to the development and organization of cell wall in filamentous fungi. Therefore, as an example, cell wall of *Neurospora crassa* is thin at the apical regions (around 50 nm) while the regions behind the apex are much thicker (around 125 nm). Cell membrane or the plasma membrane of the cells of the fungi is composed of a bilayer of phospholipids with globular proteins interspersed within them (Figure 1.2). The role of these proteins is to control movement of nutrients and metabolites into or out of cells. It also acts as a selective barrier for their transportation. Some types of sterol are also present within the fungal membranes similar to plants and animals. However, on contrary to cholesterol in animals and phytosterols in plants, the major type of sterol found in fungi is ergosterol. The exploitation of this difference is done during the treatment of fungal infections. When antifungal agents are used, the sterols are used for quantification of fungal growth.

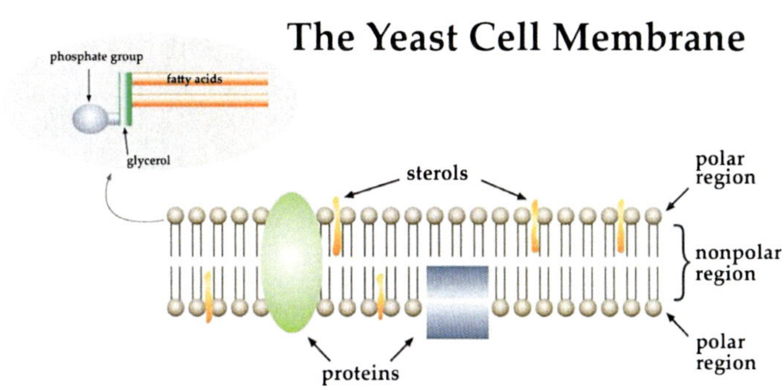

Figure 1.2: The yeast cell membrane. (Source: *http://distillique.co.za/catalog/images/articles/yeast_cell_membrane.jpg*).

The region between the cell membrane and the cell wall is termed as the periplasm or periplasmic space. Within this region of yeast cells, proteins that cannot go through the cell wall reside. These include several secretory proteins, such as mannoproteins, and enzymes, such as invertase and acid phosphatase. In filamentous fungi, the cell membrane and wall may be intimately bound as hyphae and are often resistant to plasmolysis (Figure 1.3).

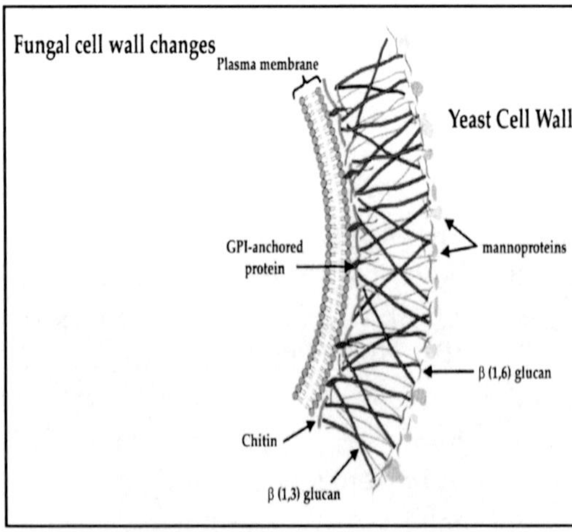

Figure 1.3: The yeast cell wall (Source: *http://www.nammex.com/wp-content/uploads/2015/02/2-fungal-cell-wall-2.jpg*).

The surface structures present on fungal cells can be seen using a scanning electron microscope. However, a more beneficial microscopy is achieved through the use of atomic force microscopy which allows visualization at a resolution of nanometers. The latter type has certain advantages such as, it doesn't require the cells to be fixed or killed. In addition, it excludes the development of potentially misleading artifacts that arise while cells are prepared for electron microscopy.

The cell wall of fungi consists of a complex network of fibrils. In filamentous fungi, variations are observed in the polysaccharide compositions depending on the taxonomic group. Along with a variety of proteins and glycoproteins, the cell wall may be composed of polysaccharides like chitin, chitosan polyglucuronic acid, cellulose, mannoproteins or glucans. An amorphous matrix of the cell wall generally has a network of semi-crystalline microfibrillar components embedded centrally into it. The overall structure of cell wall is strengthened by the bonding that occurs behind the extending hyphal tip. Several evidences revealed the dynamic nature of cell wall. Studies have shown that there are several quantitative and qualitative variations among the different fungal species, and also among the various morphological forms of a particular species. These variations are also seen to be induced as a result of environmental stress. For instance, certain fungi produce hydrophobins, a type of hydrophobic proteins within their appressoria (structures in pathogenic fungi that aid in penetration into host) or aerial growth. In another case, biotic and abiotic stresses induce formation of the pigmentation melanin in some fungal species that provides protection.

In higher fungi, the development of hyphae occurs through tip growth which is followed by formation of cross-walls (referred as septation). On the contrary, the lower fungi tend to remain aseptate except in a few cases. (Figure 1.4) Hyphae are provided with structural support through formation of septa which compartmentalizes the hyphae. Frequently, the septa are perforated in order to allow the cytoplasm or even protoplasm between compartments to move and communicate. These septal pores can become blocked by structures like Woronin bodies. This is advantageous for differentiation of the fungal cells, both morphological and biochemical. It also aids in blocking off hyphae that are stressed or damaged from those that are healthy and undamaged. Types of pores vary among different taxonomic groups and species.

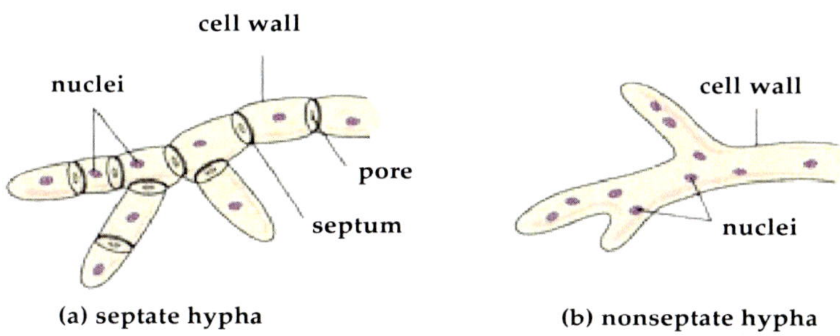

Figure 1.4: Types of hyphae. A) Septate hypha B)Aseptate hypha, (Source: *http://www.medical-labs.net/fungi-hyphae-1522/forms-of-hyphae-septate-and-coenocytic-hyphae/*).

The structure of yeast's cell wall is composed of polysaccharides, mainly β glucans, which provide rigidity to the structure. Additionally, it contains proteins, predominantly mannoproteins on the outer layer of cell wall which determine porosity, along with lipids, chitins, and inorganic phosphate. Cell walls of the hyphae consists of fewer mannan compared to yeast cell forms. Such compositional changes are also seen when unicellular fungal form is transitioned into growth of mycelium in dimorphic fungi (Figure 1.3).

Bud scars in yeast cell wall are made up of chitin (Figure 1.5). These scars are formed as remnants on the mother cells after the daughter cells bud from them. These bud scars can be stained using fluorescent dyes such as calcofluor white. This staining of scars helps to study various aspects of yeast cells including their age. The number of scars formed represents the number of cell divisions that the cell has gone through during its life.

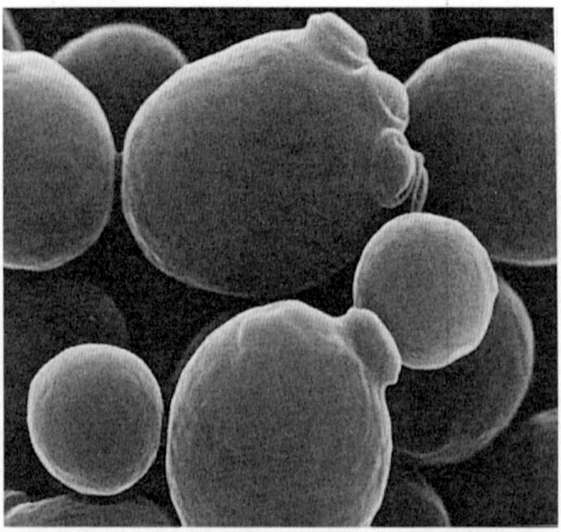

Figure 1.5: An SEM image of *Saccharomyces* showing a developing bud and bud scars. (Source: *Walker, G. M., & White, N. A. (2005). Introduction to Fungal Physiology Fungi (pp. 1–34): John Wiley & Sons, Ltd.*).

The fungal cell wall is often covered with extramural layers such as fimbriae or capsules.

Certain *basidiomycetous* and *ascomycetous* fungi have protrusions extending from their cell walls that aid in cell-cell interactions. These are referred to as fimbriae and they are composed of long proteins. In addition, some *basidiomycetous* fungi also contain polysaccharide-rich structures that are called as capsules. These capsules help to protect the fungal cell in stressful conditions. In some cases, presence of capsule may indicate the yeast's virulence properties and evasion from macrophages, for example in *Cryptococcus neoformans.* Another extrahyphal substance called as pullulan is a polymer that is commercially produced from *A. pullulans*.

1.3.2. Cellular Structure and Organelle Function

The fine intracellular structures of a fungal cell can be observed using a transmission electron microscope. This can be seen in figure 1.6. An aqueous solution called as the cytoplasm has dissolved salts, metabolites, and proteins in it. The cytoplasm consists of subcellular compartments called as organelles. As the cells of the filamentous fungi become older within the central colonies, they lose their protoplasm and organelles. The components of the protoplasm are utilized and recycled to allow growth of hyphal tips that contain actively growing cells. Additionally, there are microbodies, ribosomes, proteasomes, lipid particles and a cytoskeletal network. The cytoskeleton is a structure that provides support and stability to the cytoplasm of the fungi. It is made up of microfilaments and microtubules. Other fungal organelles that are membrane-bound include nucleus, endoplasmic reticulum (ER), mitochondria, Golgi apparatus, secretory vesicles, and vacuoles. A membranous system is extended from several of these organelles. As an example, the membrane of the endoplasmic reticulum is contiguous with the nuclear membrane. Secretory proteins of the fungi involve transportation via membranes of ER, Golgi apparatus, plasma membrane, and vesicles. The image of a typical yeast cell with some of its organelles are visible in Figure 1.6.

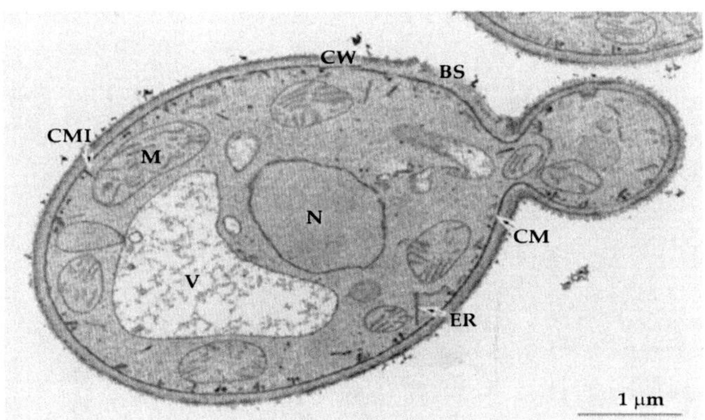

Figure 1.6: A typical yeast cell as seen under an electron microscope.CW: Cell wall, CM: Cell membrane, CMI: Cell membrane invagination, BS: Bud scar, M: Mitochondria, N: Nucleus, V: Vacuole, ER: Endoplasmic reticulum. (Source: *Walker, G. M., & White, N. A. (2005). Introduction to Fungal Physiology Fungi (pp. 1–34): John Wiley & Sons, Ltd.*).

The fungal cells are eukaryotic in nature due to the presence of a nucleus. This organelle, like other eukaryotes, is bound by a double membrane which encloses the genetic material immersed within a nucleoplasm. Majority of the yeast and fungi are haploids while some may switch between diploidy and haploidy during their lives, for example, *S. Cerevisiae*. Fungal chromosomes constitute of DNA stabilized by proteins. During mitosis, they duplicate and divide among newly formed cells or compartments of hyphae. This segregation ensures that after cell division each of the daughter cells and the compartments of the septa formed receive the genetic material. While yeast cells are made up of a single nucleus per cell, the septa of filamentous fungi may be made up of one or more nuclei. Each hyphal compartments of Monokaryotic basidiomycetes possesses one nucleus, whereas two or more genetically different haploid nuclei are observed in dikaryons or heterokaryons. Due to the presence of multiple nuclei in each hyphal compartment, the fungi are able to exhibit both haploidy and diploidy. A further discussion on this will be done in Chapter 2.

A region called as Spitzenkorper is often observed within the apex of hyphal tip growth in filamentous fungi. It is also referred to as the apical vesicle cluster or center or apical body. This region is observed as a phase-dark near-spherical region that consists of clusters of small membrane-bound vesicles. It can be stained with hematoxylin and is visible under a light microscope. Emerging from the vesicle-free core of this region are microfilaments and microtubules. Various sized vesicles are present within Spitzenkorper that are formed from Golgi bodies. Both the larger vesicles and the microvesicles (referred as chitosomes) vary in their compositions. During the growth of the tip, it positions itself to the side when the growth direction varies and eventually disappears when the growth stops. This vesicle center promotes extension of fungal wall hence facilitating apical growth, branching, clamp connection formation (in Basidiomycetes) and germ tube formation.

1.4. NUTRITION IN FUNGAL CELL

1.4.1. Chemicals Required for Growth of Fungi

Fungal cells and yeasts are heterotrophic and have relatively simple nutrient requirements. They are able to thrive in aerobic conditions with sufficient supply of glucose, ammonium salts, inorganic ions and some growth factors. The symbiotic species, like vesicular-arbuscular mycorrhizal (VAM) fungi, are exceptional cases that survive only when cultivated in the presence of a plant partner. To allow ideal growth of fungal cells, certain macronutrients are needed which include carbon, nitrogen, oxygen, sulfur, phosphorus, potassium, and magnesium. These are to be supplied at concentrations in millimolar. Additionally, micronutrients needed at micromolar concentrations include calcium, copper, iron, manganese, and zinc. Some fungi are oligotrophic in nature. This means that they can survive and grow in a very limited supply of nutrients and are able to obtain nutrients in small quantities through volatile organic compounds in the atmosphere. Chemical nutritional requirements of fungal cells for their growth are summarized in Table 1.2.

Table 1.2: Elements Required for Growth of Fungal Cells

Element	Common Sources	Cellular Function
Carbon	Sugars	Combined with h oxygen, nitrogen, and hydrogen to form structures of cells of fungi. Source of energy
Hydrogen	Protons from acidic Environments	Provides the proton motive force necessary for nutrition in fungi. Maintains acidic pH inside cells which helps in metabolic reactions in fungi
Oxygen	Air, O2	Reactant for respiration and other oxidation reactions. Synthesis of ergosterol and unsaturated fatty acids
Nitrogen	NH_4 + salts, urea, amino acids	Amino acids in enzymes and proteins are structurally and functionally dependent on it
Phosphorus	Phosphates	Structure of membrane, nucleic acid, and transduction of energy
Potassium	K^+ salts	Activity of enzymes, balance of ins
Magnesium	Mg^{2+} salts	Structure of cells, organelles, and activity of enzymes
Sulfur	Sulfates, methionine	Vitamins and amino acids
Calcium	Ca^{2+} salts	Second messengers in signal transduction
Copper	Cupric salts	Redox pigments
Iron	Ferric salts. Fe^{3+} is chelated by siderophores and released as Fe^{2+} within the cell	Cytochromes and Haem proteins
Manganese	Mn^{2+} salts	Activity of enzymes
Zinc	Zn^{2+} salts	Activity of enzymes
Nickel	Ni^{2+} salts	Activity of urease
Molybdenum	Na_2MoO_4	Metabolism of nitrates and vitamin B_{12}

Source: Walker, G. M., & White, N. A. (2005). Introduction to Fungal Physiology Fungi: John Wiley & Sons, Ltd.

Some fungi are able to obtain energy and carbon by oxidation of reduced forms of organic compounds. These are termed as chemoorganotrophs. The carbon sources which can be utilized by such fungi commonly include hexoses, and polysaccharides, for example, glucose, starch, and cellulose. However, sometimes fungi can also utilize aromatic hydrocarbons for their growth. For example, white-rot fungi are able to grow on lignin. Other than these, a variety of other carbon sources that the yeasts and filamentous fungi are able to use for their growth include pentose sugars, disaccharides, trisaccharides, oligosaccharides, some aliphatic alcohols, sugar alcohols, organic acids etc.

All fungi need to supplied with a nitrogen source as they are unable to fix nitrogen themselves, i.e., they are non-diazotrophs. They can be supplemented with both inorganic forms (e.g., ammonium ions) and organic forms (e.g., amino acids) of nitrogen. The most prevalently used nitrogen source is ammonium sulfate as it also provides the fungal cells with utilizable form of sulfur. Inorganic nitrate and nitrite salts are also used to allow growth of some fungi. This needs the presence of some enzymes including nitrate reductase and nitrite reductase that collectively convert nitrates into ammonia. Nitrogen sources in the form of amino acids, amines, and amides can be assimilated by majority of fungal species. Interestingly, many of the fungi have proteolytic properties, i.e., they are able to secrete extracellular proteases that hydrolyze proteins to release amino acids that are then utilized for growth. Fungi that are able to obtain nitrogen from urea are called as urease positive. Several fungi including some basidiomycetous yeasts are included in this category. Ascomycetous yeasts, on the contrary, are unable to do so.

Most of the fungi cannot survive without a supply of oxygen and are therefore called as obligate aerobes. However, they are usually microaerophilic which means that they are prefer an environment with oxygen levels below the atmospheric levels. A few fungi are facultative anaerobes. Including *S. Cerevisiae*, these yeasts cannot live completely in the absence of oxygen. However, they are able to survive without oxygen only if supplemented with a source of fatty acids and sterols as these compounds cannot be synthesized in the absence of oxygen. Yeast species are rarely considered as obligate anaerobes. In aerobic respiration in yeast and fungi, an electron transport chain is present where oxygen acts as a terminal electron acceptor and finally produces water. Conclusively, the oxygen responses of fungi to oxygen varies from species to species and these summarized in Table 1.3.

Table 1.3: Categories of Fungi on the Basis of Metabolic Responses to Oxygen Availability

Mode of metabolism of energy	**Examples**	**Description**
Obligate fermentative	Yeasts: *Candida pintolopesii* (*Saccharomyces telluris*) Fungi: facultative and obligate anaerobes	Naturally occurring respiratory-deficient yeasts. Only ferment, even in presence of oxygen No oxygen requirement for these fungi. Two categories exist with respect to the effects of air: facultative anaerobes (e.g. *Aqualinderella* and *Blastocladia*) and obligate anaerobes (e.g., *Neocallimastix*)
Facultatively fermentative		
Crabtree-positive	*Saccharomyces cerevisiae*	predominantly ferment high sugar-containing media in the presence of oxygen
Crabtree-negative	*Candida utilis*	do not form ethanol under aerobic conditions and cannot grow anaerobically
Non-fermentative	Yeasts: *Rhodotorula rubra* Fungi: *Phycomyces*	do not produce ethanol, either in the presence or absence of oxygen Oxygen essential for such fungi
Obligate aerobes	*Gaemannomyces graminis* (the take-all fungus)	The growth of these is markedly reduced if oxygen partial pressure falls below normal atmospheric

Source: Walker, G. M. (1998). Yeast Physiology and Biotechnology, John Wiley & Sons, Ltd, Chichester. Deacon, J. (2005) Fungal Biology, 4th edn, Wiley-Blackwell Publishing.

Carlile, M. J., Watkinson, S. C. and Gooday, G. W. (2001). The Fungi, 2nd edn, Academic Press, London.

Fungi are supplied with inorganic sources of sulfur such as sulfates, sulfites, thiosulfates or inorganic sources such as amino acid methionine and glutathione. The most oxidized form of inorganic sulfur, sulfate, is synthesized by literally all of the yeasts studied.

Integral parts of fungal cells, like nucleic acids, phospholipids, ATP, and glycophosphates, are constituted of phosphorus. A considerable amount of phosphate is therefore found in fungal cells. For example, 3 to 5 % of the dry weight of some yeast cells is made up of phosphorus, majorly in the form of orthophosphate ($H_2PO_4^-$). Phosphates can be stored in fungal vacuoles as complex inorganic polyphosphates. These complexes form volutin granules. In nature, the availability of nitrogen and phosphorus may limit the growth of fungi. As a result of this lack of availability of phosphorus within the environment, some filamentous fungi have developed some biochemical and morphological strategies to obtain these nutrients. This efficiency is exploited by plants through their symbiotic relationship between their roots and mycorrhizal fungi. In plants, phosphorus is mainly stored as phytic acid. This form cannot be utilized by monogastric animals and therefore, phytases that are produced from yeast and fungi are commercially produced for reduction of phytic acid in foods and feeds.

To allow fungal growth, several trace elements including potassium and magnesium are required. The macroelements, potassium, and magnesium, that are required in millimolar concentrations, function as co-factors for enzymes. Other trace elements, called as microelements, are required in micromolar range. A variety of trace elements and their approximate concentrations that promote fungal growth are shown in Table 1.4.

Table 1.4: Metals Required for Growth and Metabolism of Fungi. a Figures are Related to Growth Stimulation of Yeast (*S. Cerevisiae*) but are Generally Applicable to Growth of Fungi.

Metal ion	Concentration supplied in growth medium’	Main cellular functions
Macroelements		
K	2–4 mM	Osmoregulation, enzyme activity
Mg	2–4 mM	Enzyme activity, cell division
Microelements		
Mn	2–4 µM	Enzyme cofactor
Ca	<µM	Second messenger, yeast flocculation
Cu	1.5 µM	Redox pigments
Fe	1–3 µm	Haern-proteins, cytochromes
Zn	4–8 µm	Enzyme activity, protein structure
Ni	^-10 µm	Urease activity
Mo	1.5 µm	Nitrate metabolism, vitamin B_{12}
Co	0.1 µM	Cobalamin, coenzymes

Source: Walker, G.M. (2004) Metals in yeast fermentation processes. Advances in Applied Microbiology, 54, 197–229.

Some organic compounds that function as growth factors in fungi are also required in minute amounts. These have roles in some enzymatic and structural metabolic reactions. Among this category, vitamins (e.g., thiamine, biotin), purines, pyrimidines, nucleosides, nucleotides, amino acids, fatty acids, and sterols are included. Requirement of external sources of growth factors indicates the inability of fungal cells to synthesize them. As a result, their growth is limited if not provided with these factors in the culture media. Some fungal species have the inherent ability to synthesize their own growth factors by utilizing glucose and hence their nutritional needs are relatively simple.

1.4.2. Culture Media for Fungal Cultivation

It is necessary to study nutritional requirements of fungi in order to allow their cultivation in labs as well as their optimal utilization in industrial fermentation. It is often easier to grow yeasts and fungi in the culture media that has acidic pH and has the presence of natural complexes that are rich in carbon, for example, malt extract and potato-dextrose. Depending on the regional availability, mushrooms can be grown on a variety of solid substrates. These include wheat straw, rice-straw, cotton waste, and fresh oak logs. Industrial fermentation y fungi can be achieved on media that has been optimized according to its application and production process. Some industrially used media have a complete nutritional supply for the fungi. For example, baker's yeast can be successfully cultivated on malt wort or molasses. However, some other processes may require to be supplemented additionally to meet the needs of the industrial process. Often, supplementation in the form of corn steep liquor, molasses or malt broth, with additional nutrients and growth from factors may be necessary. As an example, culture medium required for production of penicillin by *Penicillium* spp. must contain sucrose (3 g/L), corn steep liquor (100 g/L), KH_2PO_4 (1 g/L), $(NH_4)2SO_4$ (12 g/L), $CaCl_2 \cdot 2H_2O$ (0.06 g/L), peroxyacetic acid (5.7 g/L). However, other industrial processes need a well-defined fermentation medium, such as the growth of *Fusarium graminarium* for production of Quorn$_{TM}$ mycoprotein.

1.4.3. Nutrient Uptake and Assimilation

A wide range of nutrients are utilized by fungi and to obtain these nutrients, they employ various strategies. As mentioned earlier, these organisms are chemoorganotrophic, saprophytic, and non-motile. Depending on the availability of nutrients in their surroundings, these fungi interact dynamically with their nutritional environment by exemplifying morphological changes within them. For example, when some eucarpic filamentous fungi grow on agar, they form extensions at the periphery of their colonies in order to explore for nutrients. The dynamic metabolism of yeast is also observed in the case when they secrete enzymes for degradation of starch even though they normally do not produce hydrolytic enzymes like higher fungi.

Several barriers for the uptake of nutrients are present on the fungal cell surfaces, namely; capsule, cell wall, periplasm, and cell membrane. Fungal cell wall is not a completely porous structure. Generally, it allows molecules with an approximate molecular mass in the ranges of 300 Da – 700 Da to move through its pores. Molecules that are soluble, like monosaccharides and amino acids, are typically more utilizable for fungi.

The dictation of molecules' and metabolites' entry and exit from the cells of fungi is carried out by the plasma membrane which is selectively permeable. The rates of metabolism, growth, and division are fundamentally based upon the mechanisms of membrane transport in fungal cells. Several physical processes of transport occur through the plasma membrane, including diffusion, facilitated diffusion, diffusion channels and active transport. In filamentous fungi, ion pumps that are located behind the apex cater the active transport of nutrients such as such as sugars, amino acids, nitrates, ammonium, sulfates, and phosphates, while closer to the apex are symport proteins. Therefore, the uptake of nutrients takes place in the hyphal tips where a continuous and fresh supply of nutrients is available in its vicinity. In addition, cells located behind the apex contain mitochondria that produce ATP which facilitate the functioning of ion pumps and generate proton motive force.

1.4.4. Overview of Biosynthetic Pathways in Fungi

The biosynthesis of cellular materials takes place through the energy-consuming and reductive pathways. These anabolic reactions are mediated by dehydrogenase enzymes which chiefly utilize the redox co-factor, reduced $NADP^+$. Glycolysis which is accompanied by the hexose monophosphate pathway results in generation of NADPH. This pathway in *S. Cerevisiae* helps in degradation of around 20% total glucose. The cytosolic NADPH that is generated as a result of this pathway is subsequently utilized for biosynthesis of various fatty acids, amino acids, sugar alcohols, structural, and storage polysaccharides and secondary metabolites. Other than the NADPG, this pathway also produces nucleotides and coenzymes for nucleotide synthesis (including NAD, NADP, FAD, and FMN).

When grown on con-carbohydrate sources like ethanol, glycerol, succinate, and acetate, the fungi have ability to undergo gluconeogenesis and polysaccharide biosynthesis which involves formation of glucose from pyruvate. The process of gluconeogenesis consumes energy in the form of ATP and NADH as a reductant.

Nitrogenous compounds such as ammonium ions are used by fungi for synthesis of amino acids families. The carbon backbones of these molecules are built through precursors of intermediary carbon metabolism.

Carbohydrates are stored in fungal cells mainly as glycogen and trehalose. Like starch, glycogen is composed of β-1,4-glucan linear units and β 1,6-branches. Trehalose is disaccharide formed through a α-1,1- glucoside bonding between two α-glucose molecules.

The synthesis of glycogen takes place with the help of glycogen synthase. When nutrients become limited, glycogen can be metabolized by the use of glycogen phosphorylase. This helps in maintaining the cellular metabolism by producing energy (ATP). When glycogen is broken down in yeast cells, biosynthesis of sterols takes place in the membrane. This production is vital for application brewer's yeast and for fermentation of beer.

Trehalose, which is the other major storage carbohydrate in fungi, is synthesized from UDP-glucose and glucose-6-phosphate with the help of the enzyme, trehalose-6-phosphate synthase and subsequent conversion by phosphatase. Trehalose is not only an important carbohydrate storage form, it is also necessary for translocation of substances and in protection from stress in filamentous yeasts and fungi. These stresses may occur as a result of osmotic shock, heat shock or even as a result of host-fungal parasitic interactions.

1.4.5. Fungal Cell Wall Growth

The polysaccharides that commonly make up the cell wall of fungi are mannans, glucans, and chitin. The basic units from which these polysaccharides are made are the sugar nucleotides which are synthesized using the pyrophosphorylase enzymes. For instance,

Glucose 1-phosphate + UTP → UDP-glucose + PPi

Mannose 1-phosphate + GTP → GDP-mannose + Ppi

The glucan component of cell wall is made up of β -1,3 linkages and β -1,6 branches which are synthesized by glucan synthetases located within the plasma membrane. Chitin, which is an important component of cell wall of fungi, is made up of *N*-acetylglucosamine units. Other than its structural role, chitin is also involved in budding in yeast and in the process of dimorphism (the phenomenon by which fungi go through transition between yeast and filamentous forms). The chitin in the fungal cell wall is synthesized by addition of *N*-acetylglucosamine to a growing chain of polymer of chitin. This is catalyzed by chitin synthetases that transfer this unit from UDP-*N*-acetylglucosamine substrate. Within the Golgi bodies, the mannoproteins are first assembled in unicellular forms. They are then translocated within vesicles towards the cell wall. The growth of cell wall is facilitated by fusion of various vesicles, carrying cell-wall synthetic enzymes, wall-lytic enzymes, enzyme activators and other components of cell wall, with the plasma membrane. Their contents are then released into the tip where they together function to synthesize cell wall.

1.5. METABOLIC PROCESSES IN FUNGI

Fungi obtain their energy through the breakdown of organic compounds as they are chemoorganotrophs. Interestingly, most of the fungi are able to utilize the monomers released from the complex polymeric compounds via extracellular digestion by secreted enzymes. The enzymes are synthesized in the Golgi bodies and then transported towards the site of cell growth where they are secreted to allow extension of tips of hyphae. Once secreted, these enzymes may function while being bound to the cell wall or they may be diffused to the environment to allow decay of substrates. /to allow this degradation, several systems of enzymes are found in fungal cells including hydrolytic, oxidative, lipolytic, peroxidative, and free-radical generating enzyme systems.

The cell wall within the wood is composed of cellulose and hemicellulose components arranged to form a matrix in a 3-dimensional structure. These lignin components are digestible by only a few species of fungi that produce relevant enzymes for degradation of cellulose and hemicellulose. They contain a mixture of oxidative, peroxidative, and hydrogen-peroxide-generating enzymes that cause decoloration of the delignified wood. Such fungi are termed as white rot fungi and have certain industrial applications. For example, they are used in the paper industry, bleaching, degradation of waste for animal feeds, biofuel production through bioremediation of land

and water contaminants. The fungi categorized as brown-rot and soft-rot can only degrade the cellulose and hemicellulose part of the wood.

The process by which ATP is produced through oxidation of intermediate carbon compounds are categorized as catabolic pathways. When glucose is converted into pyruvic acid with the help of enzymes it is called as glycolysis. In fungal cells, energy is produced through this process with side products (such as NADH) that are utilized later in biosynthetic pathways. When the yeast cells undergo anaerobic fermentation, NAD+ is formed along with the pyruvate which is subsequently converted through enzymatic actions into ethanol. NAD+ is reused in the process of glycolysis. In addition to ethanol, some other fermentation products are also produced, including glycerol, fusel alcohols, esters, organic acids and aldehydes which collectively help in development of flavors in wine, beer, and whiskey.

Aerobic respiration of glucose is carried out through glycolysis, Kreb's cycle, electron-transport chain, and oxidative phosphorylation. Other than glucose, certain pentose sugars, sugar alcohols, organic acids, aliphatic alcohols, hydrocarbons, and aromatic compounds can also act as substrates for aerobic respiration. Fats are lysed extracellularly by the help of some enzymes, releasing fatty acids which are eventually metabolized through oxidation in mitochondria.

In both yeast and filamentous fungi, aerobic respiration of glucose occurs in mitochondria via a series of reactions and pathways resulting in complete oxidation of pyruvate into carbon dioxide, NADH, FADH2, GTP, and hydrogen ions. In the citric acid cycle, hydrogen atoms reduce the co-enzymes NAD+ and FAD molecules which are then re-oxidized through the electron transport chain present in the mitochondrial membrane. Transfer of electron releases energy that is used for production of ATP by oxidative phosphorylation. This synthesis of ATP is driven by proton motive force generated within the inner mitochondrial membrane. Therefore, oxidation of molecules occurs in the mitochondria releasing reduced cofactors which are subsequently oxidized by the electron-transport chain.

Fungi are able to utilize nitrogen-containing compounds to form amino acids and proteins. For example, biosynthesis of amino acids glutamate and glutamine can be carried out directly by employing ammonium ions. Breakdown of proteins can be carried out with the help of extracellular protease enzymes. In many important pathways involving synthesis of several macromolecules, glutamine synthetase plays a significant role. The amino acid glutamate is necessary for metabolism of carbon and nitrogen as well. Some other crucial enzymes for nitrogen metabolism are glutamate dehydrogenase and glutamate synthase. Fungi are able to sequester nitrogen by assimilation of ammonia into several amino acids and intermediates of Kreb's cycle. Depending on the concentration of ammonium ions and intracellular amino acids, the pathway for ammonium assimilation may vary. In some yeasts and fungi, presence of nitrate and nitrite reductases enables utilization of nitrates as the nitrogen source. Ammonium ions that are formed are subsequently used for assimilation into products like glutamate and glutamine. In some cases, urea can be utilized through conversion into ammonium ions by the enzymes urea aminohydrolase. Amino acids can be used for formation of proteins or they can be broken down through the processes deamination, decarboxylation, transamination, and fermentation. The products formed may be ammonium ions or glutamate. In addition, fermentation yields higher alcohols or *fusel oils*, such as isobutanol and isopentanol. These compounds are vital for flavoring of fermented beverages.

1.6. FUNGAL GROWTH AND THE PHYSICAL PARAMETERS

Fungi grow in environments that are warm, have acidic conditions and have sufficient oxygen. Even though fungi are able to thrive in a wide range of temperatures, most of the species prefer temperatures around 25°C. However, some fungi are psychrophilic and thermophilic as well. The ability of fungi to survive in a diverse range of temperatures depends on multiple factors, including genetic factors, physical growth parameters, and nutrient availability. Fungi do not have the ability to regulate their internal environmental temperatures. When exposed to high heat, cellular damage occurs and cell death increases. Such high temperatures lead to disruption of bonding that stabilizes proteins which become denatured. Thermotolerance is the transient ability of fungal cells to tolerate sudden exposure to elevated temperatures up to 40°C. This thermotolerance can be introduced intrinsically after a sudden heat shock or it may be induced by preconditioning the cells to a mild heat shock (around 37°C) before exposing to a severe heat shock (50°C). Sudden shifts to elevated temperatures in fungal cells leads to the synthesis of a special category of proteins called as the heat shock proteins or the Hsps. These proteins have a number of roles including protection from high temperatures.

Fungi cannot survive without the presence of water. It is essential for the cell physiology which is affected adversely when the cells are exposed to water stress. Some species of fungi are able to survive in low water potential regions and are termed as osmotolerant or zero tolerant. Other species that are unable to thrive are called as non-osmotolerant. If exposed to a hyperosmotic shock by gradually increasing the solute concentrations in the surrounding medium, the fungal cells undergo a mild stress of water. Conversely, when exposed to a high water potential, the cells undergo hypo-osmotic shock. These short-term shocks are overcome by physiological changes such as changes in internal osmotic pressure by varying glycerol concentrations which help replace cellular water, restore cell volume and enable fungal metabolism to continue. Other solutes that aid in protection against water stress include trehalose, mannitol, and arabitol.

Most of the fungi prefer acidic conditions in their environment (pH of 4–6). However, some of them are able to survive in more alkaline or acidic pH ranges. Acidity induced by the presence of mineral acids is more preferable compared to organic acids. The reason is that organic acids tend to alter and acidify the internal environment of the fungal cells which in turn alters the cellular physiology.

Other physical parameters that influence the fungal physiology are aeration, pressure, radiation, and centrifugal force.

1.7. FUNGAL REPRODUCTION AND GROWTH

When the fungi transport and assimilate nutrients, they use it to form components of cell. This leads to an increase in biomass and subsequently to cell division or septation. Reproduction in fungi can take place most commonly by budding and sometimes by fission process. Some *ascomycetous* yeasts undergo multilateral budding process. When. *cerevisiae* reach a certain size they initiate budding and simultaneously begin synthesis of DNA. Budding takes place when the cell wall becomes weakened and the turgor pressure applies tension which results in extrusion of cytoplasm into an area enclosed by new cell wall. The previously mentioned enzymes help is synthesizing cell wall components. A ring is formed on the surface of mother cell which is made up of chitin and after cell division, this ring will eventually form the bud scar.

Schizosaccharomyces species of yeasts reproduce by the process of fission. It involves formation of septum which divides the cell into similar sized daughter cells. These divided cells then become elongated until a constant length is reached. This is then followed by mitosis. The inner cell wall grows laterally to form primary septum and it grows inwardly to form secondary septum. This transverse cleavage is termed as the cellular fission.

The cellular components of filamentous fungal cells are ambiguous. In higher filamentous fungi, the apical compartments consist of multiple nuclei, therefore, the nuclear replication occurs and it gets segregated within the newly formed septated hyphal compartments. This is referred as the duplication cycle. The hyphae in filamentous fungi also divide and branch at regions behind the extending hyphal tips or behind the septum in higher fungi. This branching enables the filamentous fungi to proliferate and fill spaces in an optimal way depending on the environmental conditions. In nutrient-rich media, branching occurs more frequently resulting in denser mycelia while less dense mycelia are formed in a nutrient-poor substratum that are more effuse to allow exploration of resources.

The cytoplasmic volume alters the rate of branching and growth of tip. Thus, the hyphal growth unit is a measure of the average length of hypha required to support hyphal tip growth. It can be used to estimate the stages of growth in fungal species. When allowed to grow on nutrient media, the fungal cells once inoculated, undergo the various phases including the lag, exponential, and stationary phases. During the lag phase, the fungal cells adapt to their physical and chemical environment. In the exponential phase, the cell mass increases exponentially by constant doubling of cell number if the conditions are optimum. This phase is followed by a period where the growth rate becomes zero and is referred as the stationary phase. During this stage, the biomass of cells remains approximately constant and if not supplied with additional nutrients, the cells may die and autolyze. This cell death can also be promoted due to accumulation of toxic waste compounds, acidic pH, increase carbon dioxide, and high temperatures. The cells in this phase undergo secondary metabolism required for cell survival. Secondary metabolites that are produced industrially, therefore, require the fungal cells to remain in stationary phase.

When cultivated in static culture media, fungi grow as pellicles. When the media is agitated, they may grow as dispersed filamentous forms or in the form of pellets. The sizes of these pellets depend on the type of species, the size of inoculum and rate of agitation. Various growth forms alter the environmental conditions which affect the physiology of fungi and eventually affect the fermentation process (Figure 1.7).

When yeast and fungi are grown industrially on stationary substratum, they have several advantages over agitated free cell suspension cultures. Immobilization of cells may be done successfully through entrapment, aggregation, containment, attachment or deposition. When attached to solid support materials, the fungal cells from a natural fungal biofilms. These biofilms have several practical applications in fermentation biotechnology and are also medically important with regard to colonization of human tissue (Figure 1.8).

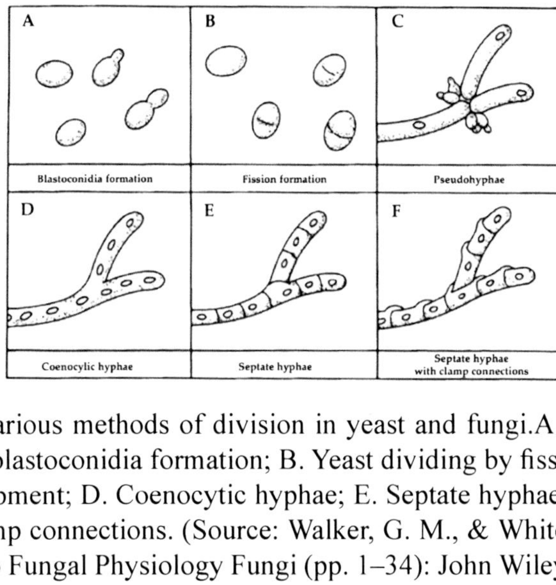

Figure 1.7: Various methods of division in yeast and fungi.A. Yeast cells reproducing by blastoconidia formation; B. Yeast dividing by fission; C. Pseudohyphal development; D. Coenocytic hyphae; E. Septate hyphae; F. Septate hyphae with clamp connections. (Source: Walker, G. M., & White, N. A. (2005). Introduction to Fungal Physiology Fungi (pp. 1–34): John Wiley & Sons, Ltd.).

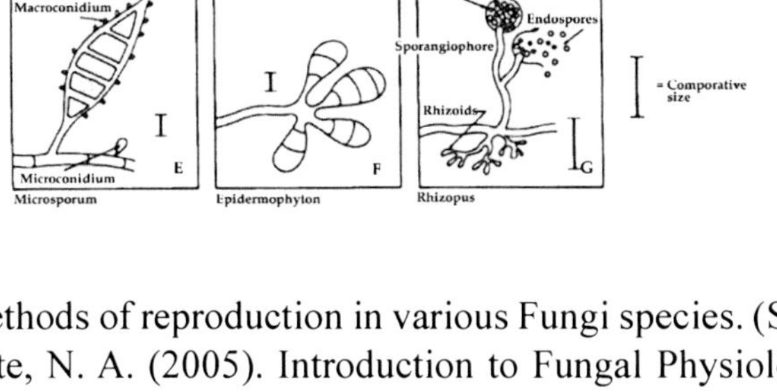

Figure 1.8. Methods of reproduction in various Fungi species. (Source: Walker, G. M., & White, N. A. (2005). Introduction to Fungal Physiology Fungi (pp. 1–34): John Wiley & Sons, Ltd.).

1.8. DEATH OF FUNGAL CELLS

Yeasts and fungi are significant systems for studying processes of aging and apoptosis. Therefore, it is necessary to study the physiology of cell death. Study of fungal cell death is important as it influences several industrial processes in fermentation biotechnology, food preservation, food production and clinical mycology.

Fungal cell death is influenced by many physical, chemical, and biological factors. It leads to a complete and irreversible failure of cell reproduction. These factors include extreme fluctuation in temperatures (both high and low), high voltage current, radiations, and high osmotic pressures. Other than these, chemical and biological fungicidal agents also result in fungal cell death. When the cells are unable to overcome these stressful conditions, physiological protection responses become weak resulting in cell damage. Industrially, physical stresses lead to elimination of fungal contaminants. This may be used for getting rid of yeast contaminants through pasteurization of foods and beverages.

Several chemical agents affect fungal growth, for example, fungicides, toxic organic compounds, free radicals and heavy metal. In foodstuff, weakly acidic antifungal agents are employed, like sorbic, aceric, and benzoic acids as preservatives. These agents are generally not able to act as fungicidal but as fungistatic agents. They function by disturbing the proton gradients within the plasma membrane and lower the pH of the cell through dissipation of ions into the yeast cytoplasm. An example of an agent that induces this acidic environment to preserve beverages from fungi is sulfur dioxide. Other fungicidal acidic agents are medium-chain fatty acids (e.g., decanoic acid), which rapidly disrupt integrity of cell membrane resulting in immediate cell death. During fermentation, certain endogenous chemicals like ethanol and acetaldehyde may also induce alterations in fungal physiology leading to cell death. These fungal cells are also killed by other factors including toxic metals, altered internal pH conditions, and oxidative damage.

Other than physical and chemical factors, some biotic interactions also lead to fungal deaths. These include ingestion, engulfment, and lysis, direct predation and intoxication. Killer yeasts are able to produce proteins that are toxic to other yeasts but they themselves are immune to them. This characteristic is exhibited by several species of yeast including *S. Cerevisiae* which produce toxins using the K1 system. The secreted K1 toxins bind to the receptors on the yeast cell wall resulting in formation of channels within the plasma membrane. This damages the permeability of the cell membrane which eventually leads to cell death. Some killer yeasts exhibit antimycotic activities which are utilized as antifungal biocontrol agents which help overcome fungal pathogens in agriculture.

Self-inflicted death may also be induced in fungi through endogenous biotic factors. These factors influence survival, physiological, morphological, genetic, and biochemical events in fungal cell. An example of this is seen in case where endogenous hydrolytic enzymes result in self-digestion of cytoplasmic proteins and polysaccharides of cell wall in fungal cells. This autolytic activity by enzymes is promoted during yeast extract production in food industry. This is done by exposing the cells to high temperatures, high salt solutions, and solvents. Exogenously secreted proteases such as papain can also lead to breakdown of fungal cell wall.

Cell death in fungi is also influenced by genetic factors. For example, fungal cells may commit suicide as a result of DNA damage in order to prevent production of a genetically defected progeny. *S. Cerevisiae* has served as a model organism for studying the processes of cellular aging and apoptosis. It has enabled the molecular studies of these two processes in eukaryotic cells. The Hayrick limit (the limit of cell division cycles) of this species of yeasts in generally 20, after which the cells enter into senescence, are unable to reproduce, and eventually dies. The aged and senescent fungal cells can be isolated and studied to isolate genes that regulate these processes. Several genes have been identified that influence stress resistance as well as longevity.

1.9. CONCLUSION

In this chapter, the physiological diversity of yeasts and fungi was highlighted and discussed in terms of morphology, growth, metabolism, and cell death. An understanding of the interactions of the fungi and their external environments is essential medical mycology in order to eliminate pathogenic fungi and also in industry to exploit these organisms for production of biotechnological commodities.

2
CHAPTER

GENETICS OF FUNGI

CONTENTS

2.1. INTRODUCTION

2.1.1. Genetic Studies Based on Fungal Model Organisms

Genetic manipulation involves the alteration of the genetic make-up of an organism in order to obtain a novel combination of traits. This has been employed extensively in the past to obtain favorable characteristics in animals, plants, and fungi. A simple method involves the identification and selection of individuals with superior characteristics, inducing variations in individuals and introducing and combining interesting traits in the next progenies. Therefore, genetic manipulations are based upon mutagenesis and genetic recombination. It can be applied for academic research to improve understanding of life and for practical purposes.

Even though initial studies on genes, inheritance, and genetic linkage were done on peas and fruit flies, the fungi served significantly as systems for elucidating fundamental insights about the genetic make-up. One of the noteworthy insights developed through utilization of fungal systems was the finding that individual genes coded for individual functions of enzymes. The genetic analysis of yeast *S. Cerevisiae* has been used for the elucidation of cell cycle and for observing the telomeres. A series of advances in unveiling numerous biological discoveries using fungal systems has proved their importance as model organisms.

Fungal genetics has been strongly influenced through the recent progress in utilizing fungi industrially on a large scale. Many achievements if genomic sciences have been made possible through studies on fungi, most significantly the yeast *S. Cerevisiae*. Remarkably, studies on fungal genome enabled the elucidation of the first chromosome sequence, first eukaryotic genome sequence and also the first production of null mutants of all the genes of an organism. Due to this pioneering work, the genomes of several species of fungi have been found and are available as annotated draft sequences.

2.1.2. Benefits of Using Fungi as Model Organisms for Genetic Studies

The genetic studies of basic cellular processes are fundamentally carried out on fungal cells. Fungal systems provide several advantages to be used as model organisms for genetic analysis. Being simple unicellular organisms, they are easier to culture as their nutritional requirements in the culture media are simple and well-defined. Obtaining mutants in the fungi is feasible as they undergo a stable haploid phase which can be easily mutated to produce organisms that have a well-characterized reproduction cycle. Additionally, fungi are eukaryotic organisms and thus are better representatives of cellular processes in human eukaryotic cells compared to prokaryotic bacterial systems that were previously employed. Genetic studies of eukaryotic cells were done using fungal systems more extensively after the progress in molecular genetics during the 1970s. Efficient genetic engineering of fungal systems is easily carried out because each individual fungal cell acts autonomously and therefore uptake of exogenous DNA results in formation of a genetically engineered complete organism. Initial success in transformation of fungi was achieved using *S. Cerevisiae* which was followed by extensive development. Subsequently, multiple vectors were constructed to allow gene transfer and cloning from chromosomes which permitted a more effective manipulation of fungal genomes. A variety of fungal strains have thus been cultured which differ from each other by as little as single nucleotide. In this chapter, we will summarize numerous aspects of fungal genetics and their employment as model organisms for genetic studies. The advantages of using specific systems within the fungal kingdom will be compared and highlighted.

2.2. LIFE CYCLE OF FUNGI

2.2.1. Saccharomyces Cerevisiae

One of the most well-studied organisms is *S. Cerevisiae* with regards to their life cycle which is also manipulable. There are two haploid forms of yeast, each representing a distinct mating type, and these can reproduce through mitotic division. Yeasts also exist in diploid forms which can either undergo vegetative propagation or through meiotic division depending on the composition of the nutrient media. The molecular pathways involved in mitosis, meiosis, cell recognition, mating, and sporulation are all well-studied at molecular level.

Mitotic division in yeasts takes place through budding which involves cell growth to be induced on a specific site on the surface of a mother cell which eventually leads to formation of a new cell (Figure 2.1). The newly formed daughter cells exhibit polarity during their growth which exploits the cytoskeletal network (made up of both actin and microtubules). It is closely synchronized with the cell cycle which ensures that each daughter cell obtains a complete set of chromosomes. Both haploid and diploid yeast cells are able to divide through the budding process with slight variations. Interestingly, some diploid cells have the ability to modify this mechanism and switch to a pseudohyphal growth pattern. Unlike the true budding mode, the fungal cells in this phase become elongated and the budding process results in building of chains of daughter cells instead of dense individual colonies.

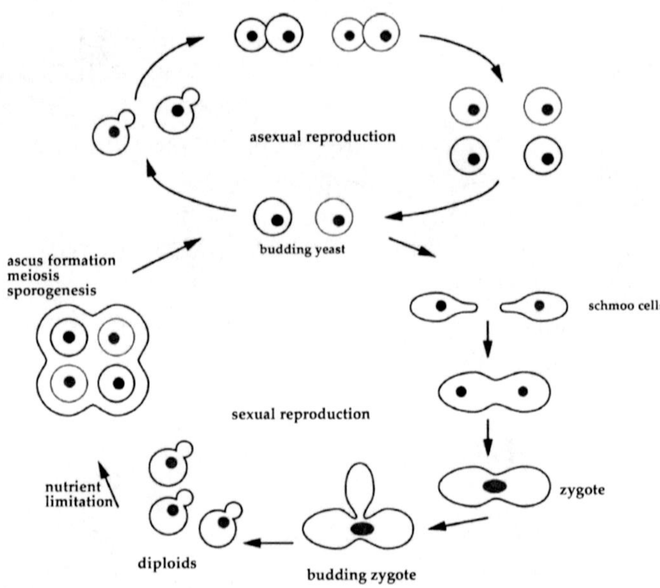

Figure 2.1: Reproduction in *Saccharomyces cerevisiae* asexually and sexually. (Source: Walker, G. M., & White, N. A. (2005). Introduction to Fungal Physiology *Fungi* (pp. 1–34): John Wiley & Sons, Ltd.).

During vegetative growth of haploid cells of *S. Cerevisiae*, the two mates come in close proximity to each other and release pheromones which results in synchronization of their life cycles and fusion of their nuclei. As a results, diploids are formed which are non-mating and carry out meiosis. The developed diploids are selected visually by micromanipulation or through observation of specific characteristics exhibited only by the diploid forms of cells. The diploid yeast cells can propagate either vegetatively in a nutrient-rich environment or through meiosis in a limited supply of nitrogen and carbon. In this way, a single diploid mating product can be amplified via meiosis.

2.2.2. Aspergillus nidulans and Neurospora crassa

As mentioned earlier in Chapter 1, filamentous fungi grow via formation of hyphae that outgrow at the tips. The branches of hyphae from the sub-apical regions initiate new tips resulting in the formation of mycelium. The hyphae are divided into compartments via cross-walls called as septa and each hyphal cell is multinucleated. Pores in the septa connect adjacent hyphal cytoplasm and hence help in obtaining and exploring nutrients from the environment. This process is carried out by enzymatic activity at the tip of the hyphae. Spores are formed as a product of both sexual and asexual reproduction. These spores can be induced to grow hyphae in filamentous fungi (Figure 2.2).

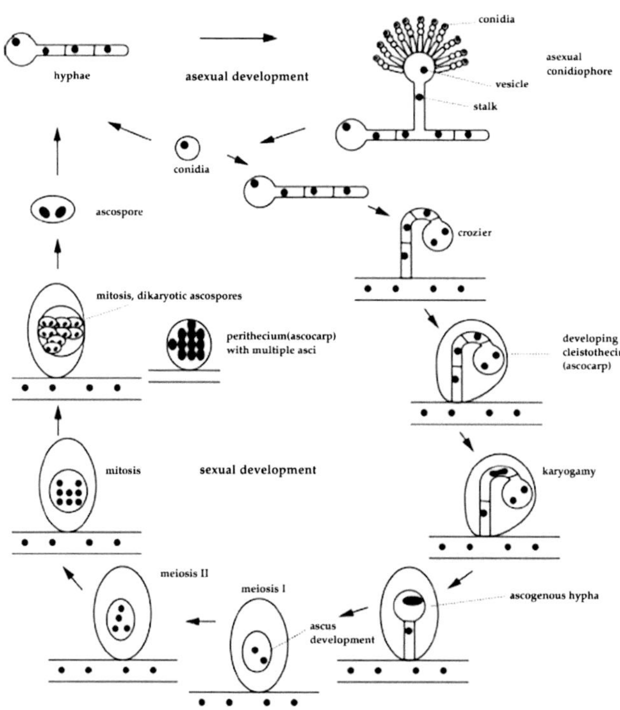

Figure 2.2: Reproduction in *Aspergillus nidulans* asexually and sexually. (Source: Walker, G. M., & White, N. A. (2005). Introduction to Fungal Physiology Fungi (pp. 1–34): John Wiley & Sons, Ltd.).

In *A. nidulans,* asexual reproduction takes place by germination of a single nucleated spore into a multinucleated hypha which eventually propagates to form branches. After some time period, a specialized branch, conidiophore, develops whose formation serves as a model developmental process. Within the conidiophore, nucleus divides by mitosis resulting in the production of haploid conidia. When released, these conidia propagate vegetatively to form hyphae and the cycle continues. Conidiation tends to occur in liquid media. It can take place on solid media which has an air interphase. Sexual reproduction in the same species occurs when two vegetatively grown hyphae fuse to form a dikaryotic hypha. Subsequently, this hypha differentiates to develop cleistothecia, a fruiting body. This structure is a mixture of different cell types including both fertile and infertile hyphae. Eventually, fertile hyphae develop into asci. Fusion of nuclei in croziers produces a diploid which undergoes meiosis to give rise to haploid spores. Upon mitotic division, haploid ascospores become binucleated and can eventually germinate into hyphae.

Aspergillus species of fungi are self-fertile, they reproduce sexually within a single colony consisting of identical genomes in nuclei. Individual strains differentiate into a cleistothecium which then develops into a crozier and eventually an ascogenous hypha. The hypha in *A. nidulans* can also undergo parasexual cycle in the form of diploids.

N. crassa undergoes asexual reproduction which produces macroconidia from the aerial parts of the hyphae which are used for subculturing. They grow vegetatively and also function in sexual reproduction. Microconidia are also produced in asexual reproductions which are relatively poor in growth and viability. Changes in light, temperature, and lack of nitrogen in media lead to a switch to sexual reproduction in *N. crassa.* In this species, mating can occur between two genetically distinct individuals. The macroconidia acts as the male partner and releases a pheromone. In response, the female forms proto perithecia, the fruiting body, which outgrows a structure called trichogyne. This structure fuses with conidia produced by male mate. Nucleus from the male cell moves into proto perithecium via the trichogyne resulting in formation of perithecium. The dikaryotic ascogenous hyphae allow division of nuclei of both mates and eventually form a crozier. Within the developing ascus, nuclear fusion occurs which is followed by meiosis and then mitosis. As a result, eight spores are formed within the perithecium which can be separated by dissection (Figure 2.3).

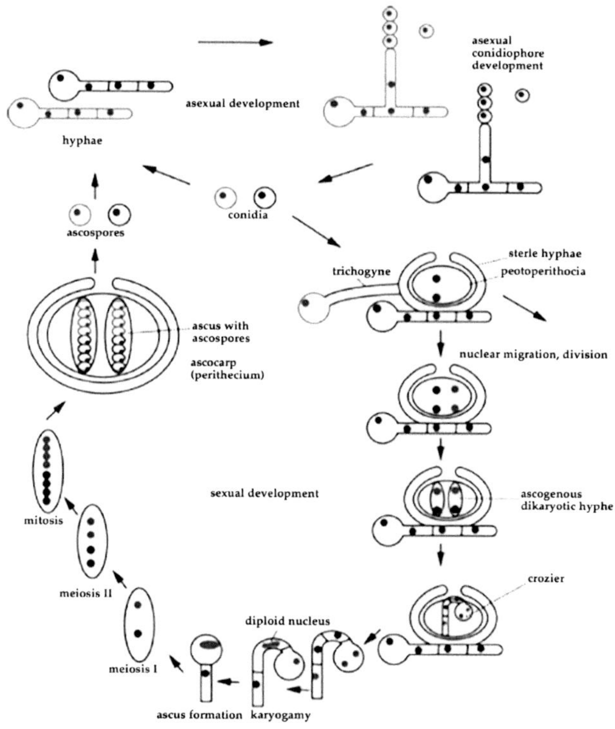

Figure 2.3: Life cycle of *Neurospora crassa.* (Source: Walker, G. M., & White, N. A. (2005). Introduction to Fungal Physiology Fungi (pp. 1–34): John Wiley & Sons, Ltd.).

2.2.3. Coprinus cinereus

The basidiomycete and ascomycetes undergo similar life cycles except for a few species. For example, the monokaryotic hyphae in *C. cinereus* produce oidia, the asexual spores, which produce new hyphae. For sexual reproduction, the monokaryotic hyphae fuse to form a dikaryotic hypha which can grow vegetatively. In basidiomycete, the hyphae can be induced to form fruiting body called basidiocarp via changes in light and temperature around it. Inside the fruiting body, nuclei fusion occurs following basidium formation. Following meiosis, haploid basidiospores are formed which are released and germinated into monokaryotic hyphae which continues the cycle (Figure 2.4).

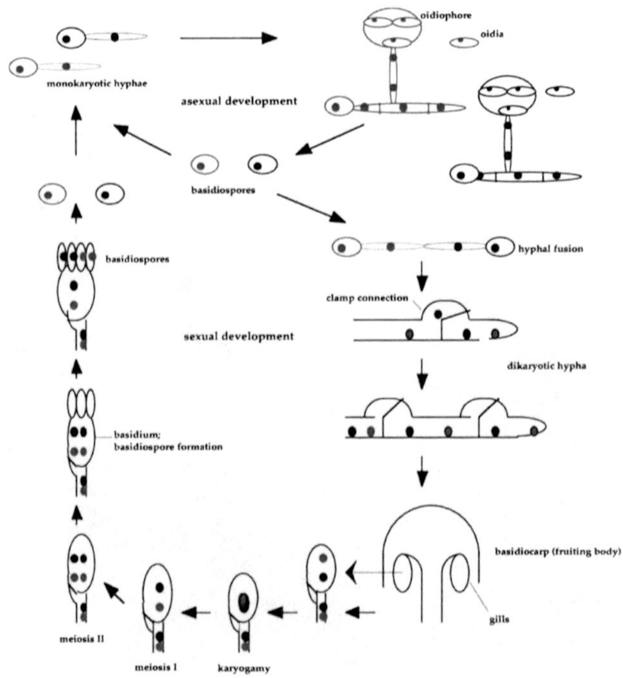

Figure 2.4: Life cycle of *Coprinus cinerus.* (Source: *Walker, G. M., & White, N. A. (2005). Introduction to Fungal Physiology Fungi (pp. 1–34): John Wiley & Sons, Ltd.*).

Cryptococcus neoformans, a human pathogen, is an example of a dimorphic basidiomycete which can take both yeast and filamentous forms. In such fungi, yeast budding results in vegetative growth. However, upon exposure to pheromones, the yeast can differentiate to form dikaryotic hyphae. At the hyphal tips, karyogamy occurs which is followed first my meiosis and then mitosis resulting in production of haploid basidiospores. These spores when released propagate as yeast. When these yeast cells are exposed to limited nitrogen or water, they go through asexual reproduction. This occurs through formation of hyphae which subsequently differentiate at the tips into monokaryotic basidia. In basidia, mitotic division results in production of haploid spores which can germinate into yeast cells.

Life cycles of filamentous fungi can vary a lot. Many of them do not undergo sexual reproduction. They can exist both as pathogenic hyphal forms and non-pathogenic yeast forms (Figure 2.5).

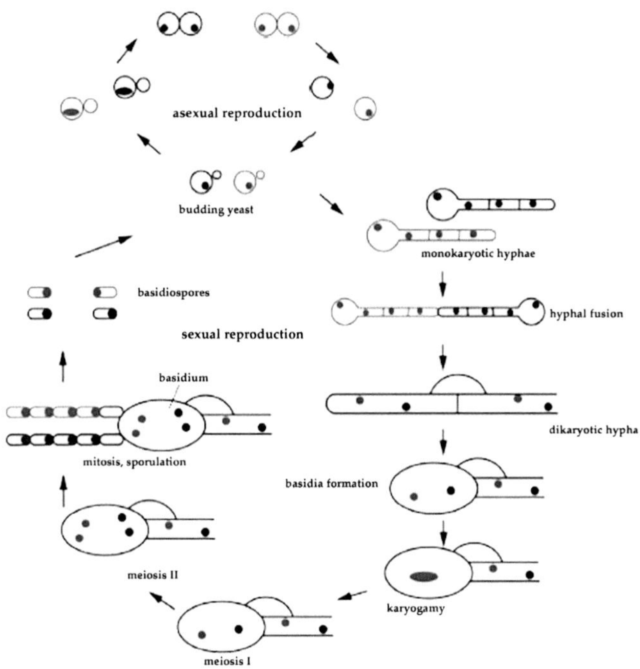

Figure 2.5: Life cycle of *Cryptococcus neoformans.* (Source: *Walker, G. M., & White, N. A. (2005). Introduction to Fungal Physiology Fungi (pp. 1–34): John Wiley & Sons, Ltd.*).

2.3. REGULATION OF MATING IN YEASTS AND FILAMENTOUS FUNGI

In *ascomycete* yeast, *S. Cerevisiae,* two haploid mating systems are found, namely, **a** and α these two mates combine to form a diploid cell. These cells cannot mate directly but they can undergo meiosis to produce four haploid cells. The mating types produced in laboratories are stable and are called as heterothallic. In nature, homothallic strains are formed which are unstable mating types. During mitosis, fungal cells can switch between the mating types resulting in formation of cells of both a andαtypes. When both the mating types mate, **a**/ αdiploid cells form. A genomic locus, homothallism (HO) controls the stabilization of mating type. This locus has a defective allele *ho* in heterothallic strains.

A single locus termed as MAT present at chromosome 3 regulates the mating type. It can either have *MAT* α allele in type a mate or *MAT* αallele in type α mate. When this locus has no alleles, mating type a is expressed automatically.

Filamentous fungi consist of mating type genes that are multi-allelic. For example, there are 12000 mating types in *C. cinerus* exhibiting a wide diversity in mating systems and complicated levels of regulation. Another filamentous fungus, *Coprinus* consists of a tetrapolar mating system. It has two distinct loci for mating type, A and B which have two types of sub-loci α andβ. These loci are redundant but recombination between them is possible. Collectively, these loci result in thousands of different mating types.

2.4. FAVORABLE FEATURES OF FILAMENTOUS FUNGI

2.4.1. Parasexual Genetics

This process has been used to map genes to chromosomes via recombination without sexual reproduction. Filamentous fungi have the ability to maintain two nuclei in a single thallus (they are heterokaryons) which makes them suitable for genetic analysis. The parasexual cycle occurs when heterokaryons undergo karyogamy (nuclei fusion) producing diploids which are recombined mitotically in a spontaneous way. Therefore, genetic recombination can take place in hyphae that are vegetatively propagating. In *Aspergillus*, heterokaryons can be distinguished from diploids due to different spore colors. When spores that are white or yellow fuse, the conidia produced can be of either color. When the heterokaryons formed are undergone karyogamy, they form diploids which have a green colored spore. This is due to complementation of recessive mutations in white or yellow spores. Once isolated, the diploid cells are induced through chemical treatment to form haploids (via loss of chromosome) which can be analyzed for crossing over that takes place during mitosis. In this stage, marker on chromosomes can be used for identifying crossing over.

In strains which do not have sexual phase involving meiosis cannot be genetically analyzed through traditional methods. Parasexual genetic analysis can be done in these strains, for example, *C. albicans* (a diploid pathogen). These strains harboring mating genes can mate to produce tetraploid products which can then be broken down to diploids by inducing chromosome loss instead of meiosis. In this strain, parasexual cycle may be used naturally for obtaining recombinations in individuals. However, the process of parasexual cycle has been used to elucidate gene linkages and in synthesis of new strains in this species. It has a promising role in analysis in future mutagenic studies.

2.4.2. Gene Silencing

In *N. crassa*, a unique feature is present, which is the RIP, repeat-induced point mutation. This feature has greatly served as a tool for molecular and genetic analysis. If multiple copies of a gene are inserted into a haploid genome in tandem, these copies are mutated (GC = AT) and inactivated during sexual reproduction. A simple process involving insertion of multiple copies of genes followed by sexual cycle, therefore, results in its inactivation.

The first eukaryotic system which exhibited a type of RNAi (RNA interference) was that of *N. crassa*. Gene silencing was achieved by inserting them at heterologous positions in the genome. This silencing is termed as quelling in fungi. It is achieved through breakdown of mRNA via several homologous protein systems involved in RNAi in other cells, including Argonaute and Dicer. To investigate the function of gene, both the process of RIP and quelling can be used together. These gene silencing studies done in *N. crassa* are representatives of similar processes occurring in other eukaryotic cells, including those of worms, humans, and plants. Interestingly, these RNAi systems can also be possibly introduced into other species of fungi which naturally lack the ability to carry out gene silencing.

2.5. GENETICS AS A TOOL

2.5.1. Tetrad Analysis

In *S. Cerevisiae*, analysis of meiosis can be done through tetrad analysis. The four spores that are generated after one meiotic cycle are present in an ascus sac. These sacs can be treated with an enzyme, endoglucanase which releases the spores. Individual spores are placed separately on culture media through micromanipulation and allowed to propagate into a spore colony. Each spore colony can be studied for the presence of markers that were initially introduced in the first cross. As a result of meiosis I, two homologous chromosomes are formed which are paired together and joined by chiasma formation. This formation induces a genetic recombination event between the two sister chromatids. Following this, the homologous chromosomes are separated and moved to opposite pole of the meiotic spindle fibers.

Nuclear division occurs which produced two diploid nuclei. At this stage, the chromosomes undergo independent assortment which is followed by meiosis II. In this stage, the attached chromatids are split and moved towards opposite poles; subsequent division of nuclei takes place which results in formation of four haploid nuclei. In yeast meiosis, all four products are available for genetic analysis. The segregation patterns need not be studied for genetic analysis. This means that a "non-Mendelian" 3:1 segregation of heterozygous markers is allowed to occur instead of 2:2 distributions. The recombinational exchange of genetic material from one allele to the other that generates this pattern is called as gene conversion. The arrangement of tetrad spores in the ascal sac is random in *S. Cerevisiae.* Identification of markers can therefore no be done on basis of position of tetrads. However, the markers that are present in centromeres can be identified on the basis of meiotic event as the centromeres segregate in the first meiotic division.

Selected tetrads is a modified approach towards tetrad analysis. Meiotic recombinations that occur within a single locus have been identified using this process. Intra-allelic recombinations are rare, therefore, selection is based on identification of recombination events that occur during meiosis. Intragenic recombination is studied using heteroalleles of auxotrophic markers. The products that are not results of recombination are auxotrophs. Unlike, tradition tetrad analysis, ascal sacs that are treated with glucanase are spread on media that lacks a particular nutrient which is needed for growth of auxotrophic cells. Any individual spore among the tetrads that initiates germination is identified microscopically and micromanipulated to allow structural analysis of genes of all tetrads.in this approach, the tetrads that contain the recombined gene of interest are identified. Thus, allowing structural analysis of the process of recombination. To develop models for studying meiotic recombinations, a detailed study of all events of meiosis is critical.

Random spore analysis is another approach that is used for experiments in which segregation pattern of marker in asci is not significant. In this process, groups of asci are together digested using glucanase enzyme. The haploid meiotic products and unsporulated diploid cells are allowed to grow on culture media. This spreading is utilized for eliminating the initial diploid cells. This is done by using a drug resistance gene as a marker which is expressed only in haploid cells. Screening of these segregated haploids is carried out by identifying a cell that consists of a combination of markers. These populations can be used for scoring the segregation patterns in the cross. A recent approach, synthetic genetic array (SGA), is a modification of random spore analysis strategy. It involves screening of complete genomes for lethal interaction which are synthetically produced by robotic replica plating. Only a particular type of haploid cell is allowed to propagate by exploiting expression of genes that are specific to mating type.

Neurospora crassa is a model organism for analysis of microbial genetics including those involving bacteria and yeast *S. Cerevisiae*. It is suitable for this purpose as it is haploid in nature and it produces asci that are larger in size allowing the recovery of all ascospores produced by meiosis to be isolated. It can, therefore, be used for studying recombination that occurs in the parental genomes of the progeny. The linear ascus in *N. crassa* has ascospores arrange linearly inside it. This allows the mapping of genes relative to the centromere positions which is particularly advantageous for the process of tetrad analysis. Following meiosis in this species, another round of mitosis occurs which produces eight haploid nuclei. When these nuclei undergo sporogenesis, they result in formation of eight ascospores located inside the ascus. Similar to *S. Cerevisiae*, the spores can be collected and tested for presence of traits like nutrient requirements.

Recombination does not happen in genes that are linked. Linkage of two genes occurs when they are located very close to each other, such that there is no chance of chiasma formation in their loci. The resultant progeny receives 50% of each of the parental type of genes at these loci due to independent assortment at meiosis I. When the two genes are far apart on the chromosome, there are greater chances of recombination. The greater the distance between the two genes, the more the proportion of recombinants that is formed, with a maximum of 50%.

Neurospora is a convenient system for tetrad analysis, as mentioned earlier, because it has an ordered arrangement of tetrads. The first division products are placed within one half of the ascus. Sexual reproduction is carried by growing the culture of female strain first and then allowing the conidia of the males strain to spread over it. After a few days, development of perithecia occurs, followed by maturation of asci which begin to release the ascospores. These spores may be collected and genetically analyzed. Alternatively, the asci may be released from the perithecium by dissecting and crushing it. The asci are poked with a needle to free a spore which allows the ascus to be broken for releasing the other spores. These spores are subsequently released and analyzed.

Before the advent of genome sequencing, the mapping of genes and linkage analysis was done by through tetrad analysis. This was done by the help of marker genes present in reference strain which is compared to the gene of interest in a test strain. The linkage of this gene with the marker gene was therefore detected and used for mapping the gene.

The species *Aspergillus nidulans* are not as easily exploited for tetrad analysis as *N. crassa*. All the unlinked genetic markers segregate by Mendelian segregation within the spores. For genetic analysis via sexual reproduction, two strains with different auxotrophy containing distinct genetic markers are grown selectively on media. This selection enables formation of heterokaryons which are allowed to propagate into fruiting bodies or cleistothecia form for 2–3 weeks. The asci are too fragile to be dissected out and hence cannot be used for tetrad analysis in traditional way. Alternatively, the cleistothecium is isolated from cellular tissue, crushed, and spread across an agar plate to propagate into spores that exhibit various segregation markers (for example, spore color). The extent of recombination and the genotype of the produced spores can be determined by growing and testing the spores on various selective media.

2.5.2. Genetic Screening through Molecular Techniques

2.5.2.1. Transformation and Plasmids

Transformation of a fungus with DNA was first carried out on *S. Cerevisiae*. With advances in technologies, the manipulation of the yeast genome became for effective. Initially, a transformed yeast strain was developed which consisted of two mutations at a single locus, making sure that the gene remained non-functional, and finally insertion of a bacterial plasmid that harbors a cloned copy of the inactivated gene. Protoplast fusion technique was employed for DNA uptake in the yeast cells. The confirmation of the presence of exogenous gene in cells having restored functions was then done through plasmid sequencing. The transformation protocols have now been modified for improving the efficiency of the experiment by treatment with lithium salts or by using electroporation. In filamentous fungi, DNA uptake can be achieved using protoplasts or asexual conidia. Calcium or polyethylene glycol is utilized for inducing DNA uptake by protoplasts. Direct DNA uptake into conidia of some fungi is done through electroporation, without formation of protoplast. Different fungal species depict varying efficiencies using different methods for transformation.

Selection of positive transformants is carried out using selectable markers, which are usually drug-resistant or nutrition genes. This type of selection is commonly used all fungal transformation systems, for example, ability to synthesize uridine and uracil.

Initial transformation experiments done in yeast did not prove to be efficient and resulted in integration of sequences into genome. The wide scope for usage of episomal DNA for bacterial transformation led the scientists to look for equivalent tools for yeast transformation. Eventually, two forms of the episodes were developed; one was developed from the backbone of yeast 2μ plasmid. While the other form harbored origins of replication from the chromosomes. The former were more stable as they consisted of genes allowing efficient segregation compared to the latter which have autonomous replication sequences (ARS) elements that are derived from chromosomes. Furthermore, to improve segregation efficiency and to maintain a low copy number of plasmid, the ARS-based plasmids can be made more stable through insertion of centromeric sequences. Such plasmids are currently being utilized extensively for genetic manipulations in yeast cells as they can be maintained for many generations without the need for selection. Plasmids used for yeast transformation are selected using nutritional markers such as LEU2, TRP1, URA3, and HIS3 which were developed by complementing mutations in *E. coli*. Drug-resistance markers are also extensively used in fungi transformant selection, for example, G418 resistance is a suitable marker in *S. Cerevisiae*. Interestingly, all yeast transformation plasmids can be easily shuttled between prokaryotic and eukaryotic systems as they usually consist of origin of replication and selection markers for both the yeast and the *E. coli*. On contrary to yeast *S. Cerevisiae*., the filamentous fungi mostly are unable to maintain extrachromosomal plasmids which typically end up integrating into the genome both homologously and heterologously. As the length of homologous DNA increases, the chances of homologous recombinations also increase.

Transformation technique can be utilized for multiple types of genomic manipulations, including gene knocking out, gene replacement with a mutant, or altering gene expression. For knocking out a gene, the 5' and 3' ends sequence of the gene of interest are cloned flanking the marker gene. When transformed, the construct becomes recombined at the 3' and 5' flanking ends with the endogenous gene resulting in its replacement with the marker gene. Expression of a gene can be regulated by insertion of a promoter beyond the endogenous ORF. The promotor for GAL1, the gene regulated by galactose is often used in *S.cerevisiae*. This usage allows either overexpression or repression of the gene in the medium containing galactose or glucose, respectively. Regulation of gene expression in *A. nidulans* is often carried out using the promotor for*al cA*, which is positively regulated by ethanol in the medium or negatively regulated in the presence of glucose.

2.5.2.2. Genetic Screening

In the construction of mutant screens, genetic analysis is still a very useful tool despite of extensive successful sequencing and annotation of fungal genomes. A vast amount of information about cellular processes has been revealed through genetic screens carried out on filamentous fungi. In addition, identification of the first conserved genes and their respective functions has been done using these organisms. Beadle and Tatum utilized *N. crassa* and demonstrated that individual enzymes were coded by individual genes, thus bringing both biochemical and genetic analysis together for the very first time. Moreover, information about recombination events, circadian rhythms, gene conversion, gene silencing and DNA methylation has all been shown using this species of fungi. Subsequent genetic screens using *A. nidulans* and *S. Cerevisiae* have helped study vast areas of research including regulation of cell cycle, cellular motors, and signaling, dynamics, and development of the cytoskeleton (Figure 2.6).

In a classical screening process, cells are first mutated through radiation, chemical or UV light and the survivors are then allowed to grow. This approach was first utilized in 1941 by Beadle and Tatum who produced *N. crassa* metabolic mutants by exposing the conidia colonies to X rays on a medium with complex composition. This was then followed by transferring to a minimal medium. Growth on both types of media were compared and the colonies that were unable to proliferate on minimal media were shown to exhibit nutritional mutation. This means that if the strain was unable to grow in the absence of a certain nutrient, it harbored mutation in the gene required for its synthesis (for example, tyrosine, leucine, or alanine). Genes for vitamin B6, B12, and *para*-aminobenzoic acid were all identified using this mutation approach. To ensure that only a single gene was mutated, the mutant strains were subsequently crossed with the parental wild-type strains.

Genes that control the circadian rhythm in the species *N. crassa* were identified using another classical mutagenesis approach. Some biological processes in fungi, as well as humans, are regulated by both temperature and light. Their regulation oscillates during the 24 hours of a day, i.e., in a circadian rhythm even without the stimuli in the environment. Such an internal clock is known to regulate formation of conidia in *N. crassa.* Mutagenesis screens and race tube assays were together used to elucidate the circadian clock regulation in this species. The race tube, which is a long glass horizontal tube, is inoculated at one end, the hyphae of *Neurospora* grow towards the opposite end building a periodic banding pattern because of the colored conidia formation every 24 h. When mutagenic conidia are inoculated, variations in banding patterns were observed due to changes in day length were screened. Through this experiment, *frequency*, *period*, and *chromo* genes were elucidated which are homologous to those present in *Drosophila melanogaster* as well as humans.

Classical genetic screens in *S. Cerevisiae* revealed several cell division cycle (CDC) genes. These discoveries helped identify significant regulatory elements taking part in the eukaryotic cell cycle. It was important to identify conditional mutations that formed a terminal phenotype, such as temperature sensitivity, in studying these set of genes as mutagenic cell cycle would result in blockage of cell growth and proliferation. This formation of a uniform terminal phenotype occurs due to absence of a critical cellular function, involved in cell cycle, at a particular temperature. This screening resulted in discovery of several key proteins important in the cell cycle, including Cdc28p, Cdc35p, Cdc12p, and Cdc9p which function as regulator of the process of mitosis and G1 to S transition, adenylyl cyclase, septin, and DNA ligase, respectively. Mutant screens helped unveiling several novel processes and identification of CDC mutant genes which are now known to be important for *S. Cerevisiae* cell biology. Since then, there have been improvements achieved in frequencies of mutant identifications by introducing several modifications in the classical genetic screen approaches.

The exploitation of mutant screens has become more prevalent, specific, and more powerful since several selection methods were introduced. Mutants in *A. nidulans* for the cell cycle process were identified through a classic screen in 1975 by Ron Morris. UV light was used to mutate conidia of this species and the mutated cells were grown at 32°C. Temperature sensitivity was then screened by replication onto media grown at 42°C. Those cells that were unable to grow at the restrictive temperature were studied further through phenotypic variations and by visualizing of the nuclei through staining. Through this study, identification characterization of some mutant classes was done, including NIM (never-in-mitosis), BIM (blocked-in-mitosis) and NUD (nuclear distribution) mutants. Complementation analysis allowed identification for the genes involved in these mutant phenotypes. BIMC (kinesin motor) and NIMA (important in G2 to M transition) were elucidated which were eventually identified both in yeast and man. Mutant screen also led to identification of a gene, NUDF, which is homologous to a human gene whose mutation leads to the human neuronal disease lissencephaly.

In *A. nidulans* another landmark was made using benomyl-resistance gene which allowed identification of mutants of tubulin. Subsequently, these *ben*A mutants were identified to be β-1 and β-2 tubulin. *Ben A* mutants were undergone suppression analysis to find out the mutations that influenced β-tubulin – interacting proteins. Studies in bacteriophage revealed that mutants of a particular protein which caused inhibition of its interaction with another protein can have their functions restored by compensatory mutation in a second factor. This type of mutation helps in restoring the ability of the two proteins to interact and hence restoring function. Therefore, a second round of mutagenesis was done on *benA* mutants which resulted in isolation of revertants that were temperature-insensitive. α-tubulin, which dimerized with β-tubulin was recognized using further analysis. *mipA* mutants were produced through subsequent suppression screening which were mapped and cloned and resulted to form an example of γ-tubulin for the first time in an organism.

In another study, *S. Cerevisiae* mutants with a non-functional mating process were used. Selection was based upon the funding that in the presence of mating pheromone produced by the opposite mating type, a functional mating response pathway caused the arrest of the cell cycle in these cells. Therefore, mutagenic *MATα*cells could only grow in the presence of the mating pheromone α-factor when they were insensitive to the pheromone. May key kinases and regulator of this pathway were identified through this approach.

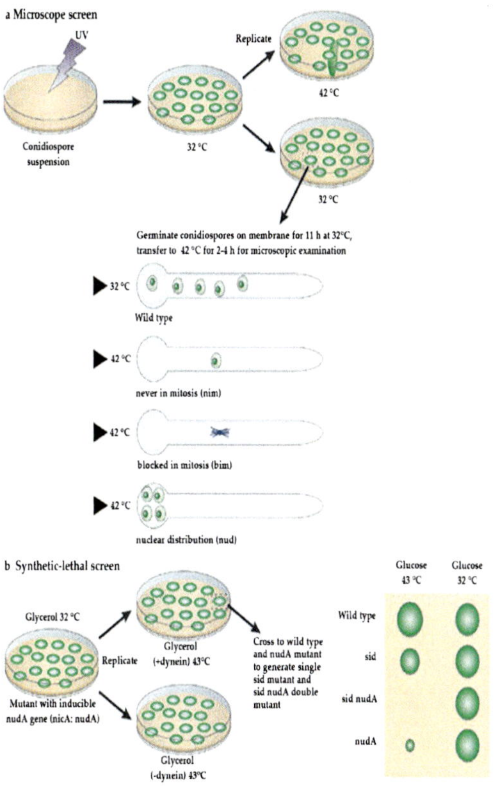

Figure 2.6: Classical genetic screening approach, (Source: *Casselton, L., & Zolan, M. (2002). The art and design of genetic screens: filamentous fungi. [10.1038/nrg889]. Nat Rev Genet, 3(9), 683–697.).*

Recently, developments in mutagenesis screens has allowed identification of synthetic lethal which can help reveal potential functionally interacting proteins. In one study, nudA(dynein) gene from *A. nidulans* was controlled by a regulatable promotor, *alcA*. Expression of nudA was possible only in the absence of glucose and in the presence of glycerol. This strain was mutated using UV light and was then grown on glycerol at 32°C. This was followed by respotting on fresh media at 42°C. Screening for temperature-sensitive mutants that were dynein positive and could grow on glycerol was done against those that were dynein negative and couldn't grow of glucose. Subsequently, isolation of single and double mutants was done by back-crossing the product strains with wild-type and *nudA* parents. Synthetic lethal with dynein (sld) were the mutants that were cloned and identified to be homologous to the spindle checkpoint factors, Bub1p and Bub3p.

In *S. Cerevisiae*, the insertional mutagenic process is highly progressive. Previously, transformation into cells was carried out using circular plasmids that harbored genes for selectable markers which got efficiently integrated into homologous sites. The mitotic and meiotic recombinations were successfully analyzed through insertion of marker genes into repetitive DNA (for example, those in ribosomal DNA). The technology of direct gene replacement was greatly improved when the double-stranded DNA breaks were recognized paving way for a more enhanced insertion of genes and targeting of homologous sites of genome in a more efficient way. Currently, there are multiple methods developed for disrupting the function of a gene of interest which part of disrupt or replace part or complete gene within the genome. Investigations of individual genes have been carried out using such gene inactivation approaches. Similar approaches have been employed in sequences of yeast that are interrupted by several random insertional mutations. A library of *E. coli* plasmid vectors has been constructed which harbor yeast DNA inserts of up to 10 kB sizes. *E. coli* is transformed using this library of plasmids and allowed to grow. These are eventually induced to undergo a transposon mutagenesis using a bacterial transposon that consists of selectable marker for the yeast *S. Cerevisiae*. This is followed by selection of plasmids that have received a transposon insertion, frequently occurring within the yeast DNA part of the plasmid. Yeast DNA inserts are isolated using restriction enzymes and are used for transformation into yeast cells. The integration events can be detected using yeast-specific markers present on the transposon. These insertions take place randomly throughout the yeast genome which depends on the randomness of the yeast sequence library and of the transposon insertions into this library. The transposon hops can be carried out using techniques that use the endogenous yeast transposable elements but these developed strategies have a nonrandom pattern of insertion into the yeast. The genetic screens in filamentous fungi have also been carried out using insertional mutagenesis approach. Restriction enzyme-mediated insertions (REMI) are done by linearization of plasmids using specific restriction enzymes, followed by transformation into protoplast in the presence of the same restriction enzyme. This enzyme aids in cleaving the host genome at specific sites which forms sticky ends compatible with the linearized DNA. Thus, multiple integrations of the plasmid into the genomic DNA may occur resulting in numerous mutagenic transformants. DNA from the transformant is isolated and cut with enzymes that do are unspecific to the plasmid DNA. This allows generation of DNA fragments that contain plasmid DNA along with the flanking genomic DNA which can subsequently be ligated and transformed into *E. coli* strains. The plasmid can be rescued using drug-resistance marker. The location of the initial plasmid integration and the identification of the mutated gene can, therefore, be achieved. In *Saccharomyces*, a more modern and efficient techniques is used for random mutagenesis. This involves introduction of a tag on the mutated DNA and through this method, a library for this strain has been developed which contains the transposon mutations. By mutating the genome of *Magneporthe griseae* using transposons, a cosmid library was developed which was termed as TAGKO. Mutagenic cosmids were sequenced through transposon integration and subsequently annotated. To study the effect of replacement of endogenous gene with a mutated gene, transformation of the cosmids containing known mutated genes into the fungi was carried out. Other filamentous fungi, like *A. nidulans*, have also been used for transposon mutagenesis (Figure 2.7).

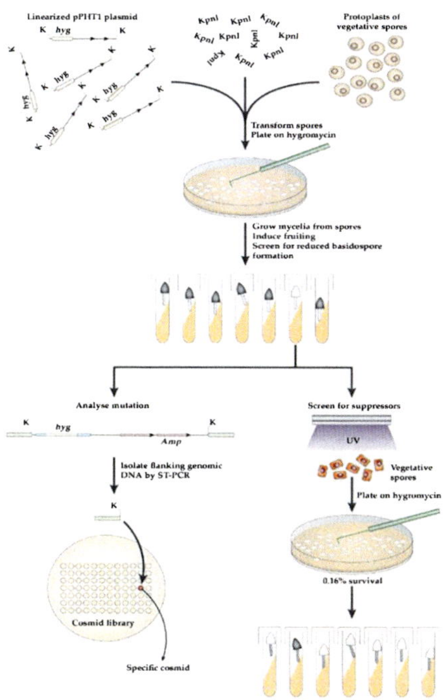

Figure 2.7: Gene cloning by insertional mutagenesis in *Coprinus cinereus.* (Source: *Casselton, L., & Zolan, M. (2002). The art and design of genetic screens: filamentous fungi. [10.1038/nrg889]. Nat Rev Genet, 3(9), 683–697*).

2.6. CONCLUSION

Fungi are important to humans in several aspects. Organisms that are economically important, like baker's yeast, *S. Cerevisiae*, are used to produce bread and alcohol. Some species, including *C. albicans* and *A. fumigatus*, are significant human pathogens. Fungi also serve as systems for investigation of molecular and biochemical processes of eukaryotic cells. Some of the most sophisticated genetic systems utilized in research and academic purposes are fungi. Through exploitation of fungi, some of the major achievements have been made in the field of eukaryotic genetics. It began from elucidation of the relationship between enzymes and genes through *Neurospora* and has been extended to the recent utilization of *S. Cerevisiae* and *S. pombe* in revealing regulation and control of cell cycle and proliferation.

Formation of novel combinations of variations is the key characteristic of genetic systems, which is highly developed in fungal systems. Fungi have especially played a leading role in the elucidation of meiotic events and DNA recombination events as they allow all of the products of meiosis to be isolated and studied. This ability allows detection of gene conversions and post-meiotic segregations.

The insights of the utilization of fungi in elucidating functions of eukaryotic cells are much deeper than what has been discussed in this chapter. Current work is being carried out using a few fungi but many more potential species of fungi may be sequenced and eventually utilized through both genetic and genomic tools. It is expected that in the future we will be able to find out more about the fungal diversity and hence unveil more fungal species that may serve as model organisms with more sophistication. The success of fungi in research during the previous years hints their extensive usefulness which may be utilized in the future research for revealing further insights about various cellular processes.

3
CHAPTER

FUNGAL GENOMICS

CONTENTS

3.1. INTRODUCTION

The study of the genome of organisms' sequence is called as genomics. Around 1300 bacterial genomes have been sequenced since 1995. In 1996, the first eukaryote that was sequenced was *Saccharomyces cerevisiae* (baker's yeast). Following this, hundreds of eukaryotic genomes have been sequenced including the human genome. This includes around 102 fungal genomes which have a relatively small size of their genome and their significance in biotechnology and their impact as pathogens (to both plants and humans). *S. Cerevisiae* which is one the most important fungi has several of its strains sequenced. Interestingly, the majority of the sequenced fungal species belong the phylum, Ascomycota, including many that are pathogenic to humans. With recent advances in technologies for genome sequencing, it is expected that a huge variety of evolutionarily, environmentally, and biotechnologically relevant organisms may be studied and utilized.

Among the eukaryotic genomes, a major contribution has been made by the fungal kingdom which has been moved to the forefront. Few of the sequenced fungal species are closely related while others have distant relationships, of up to 1 billion years ago. This means that fungi can be used to analyze evolutionary relationships in eukaryotic organisms, with reference to the content, structure, and organization of their genomes. Through the process of comparative genomics, useful information regarding the linkage between genes and their functions may be revealed. For example, comparing two or more closely related pathogenic and non-pathogenic species may help in isolation of metabolic pathways and the genes associated with virulence.

3.1.1. The Fungal Kingdom

Fungi are one of the kingdoms of life and they are eukaryotic. They are not autotrophic in nature as they lack chlorophyll. Instead, they belong to a class of organisms termed as saprobic, that feed on the dead organic matter. Initially, fungi were predicted to have a closer phylogenetic relationship with plants but recently, phylogenetic studies have revealed that these organisms are more closely related to animals rather than plants. The fungal kingdom is vast and abundant, predicted to consist of around 1.5 million species. It includes organisms that are beneficial in biotechnology, harmful pathogens, or can have mutualistic relationships with plants. According to studies related to phylogeny, Chytridiomycota, Zygomycota, Ascomycota, and Basidiomycota are the four phyla within the kingdom of fungi as shown in Figure 3.1.

Chytridmycota, also known as chytrids, is one phylum of fungi which consists of organisms that reproduce by the formation of zoospores which spread and disperse in the presence of water. The organisms in this group are ancient and are predicted to have evolved slightly after being diverged from the common eukaryotic ancestor. Chyrids have important roles as decomposers of organic matter and are mostly habitants of soil or freshwater, and sometimes marine environment. A pathogenic fungal species, *Batrachochytrium dendrobatidis*, causes chytridiomycosis (a disease in amphibians) and belong to the chytrid phylum. The phylum Zygomycota reproduce sexually by producing zygospores. The organisms of this phylum have diverse morphologies and they constitute 1% of the known species of fungi. The ecological diversity of this phylum is the most diverse, they are saprophytic on fruit, dung, soil, arthropods guts, and as pathogens of various plants, animals, and other fungi. Bread mold and fruit rots are caused by *Mucor* and *Rhizopus* species, respectively.

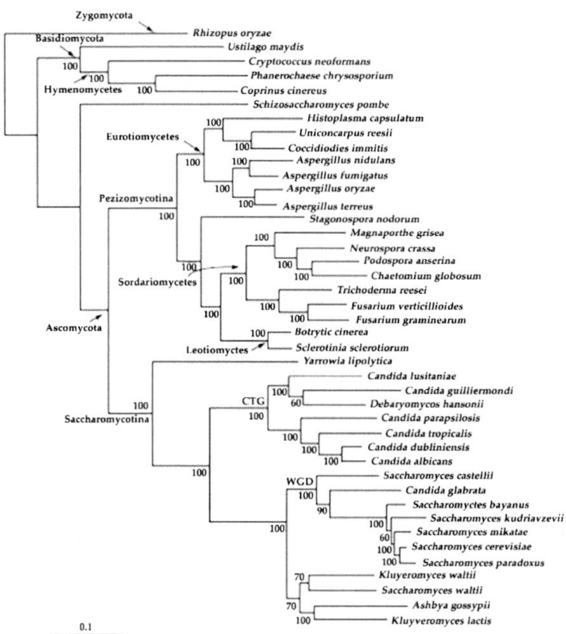

Figure 3.1: Phylogeny of fungi developed using 42 genomes. Zygomycota, Basidiomycota, and Ascomycota are three of the four phyla in fungi kingdom. (Source: *Whiteway, M., & Bachewich, C. (2005). Fungal Genetics Fungi (pp. 35–63): John Wiley & Sons, Ltd*).

About 65% of the fungal species belonging to the phylum Ascomycota. These species have an ascus, which is their distinguishing feature. These structures hold the sexual ascospores together which are produced as a result of meiosis and one round of mitosis. The ascospores formed are resistant to stressful conditions as they have thick walls. Upon germination, they propagate into haploid fungi. Ascomycota have been divided into three subphyla; Saccharomycotina, Pezizomycotina, and Taphrinomycotina. Important species like the brewer's yeast (*S. Cerevisiae*) and human pathogen (*C. albicans*) belong to the subphyla Saccharomycotina. They produce naked asci which lack an ascoma, All filamentous fungi (molds) including *Aspergillus fumigatus* (human pathogen) and *Penicillium chrysogenum* (produces penicillin antibiotic) belong to the Pezizomycotina. In Taphrinomycotina phylum, species with diverse morphologies are included. One example is the fission yeast form of *Schizosaccharomyces pombe*.

35% of all the known fungal species are part of the Basidiomycota phylum. Members of this category, such as mushrooms and puffballs, have distinct feature of large fruit bodies. The edible mushrooms including *Agaricus bisporus* (common mushroom) and *Pleurotus ostreatus* (oyster mushroom) are part of Basidiomycota phylum. A very few microbes have the ability to degrade lignin (tough material that is part of plant cell wall. Certain members of this group of fungi have this inherent ability are therefore significant species in biotechnology as they can be used for detoxification and delignification of waste products.

3.2. GENOME SEQUENCING

In axenic conditions, several microbes including fungi are unable to cultivate and propagate. However, interesting information about their occurrence, abundance, and ecological niches can be revealed through DNA-based methods. In prokaryotes, the barcoding targets are nuclear ribosomal genes and markers, especially the small subunit (SSU). Because it has universal primer sites, they are easy to amplify, and there is alignability among various phyla and domains, it is easy to be amplified. On contrary, SU genes are conserved among various species and hence do not provide species-specific information. It results in a grouping of phylogenetically close phylotypes or even genera. For fungal identification, the species-level resolution has been achieved through the employment of another exploratory barcoding target, the internal transcribed spacer (ITS) region at the 3′-end of the SSU. Information about such specific sequences of DNA among fungi help reveal useful insights about their taxa, their roles as pathogen or mutualists, and therefore, even at low relative abundance, it may influence the host performance. Several methods for DNA sequencing have been employed for fungal genomes. These have been discussed in detail below.

3.2.1. Sanger Sequencing

The first protein that was sequenced was insulin, in 1958, by Fredrick Sanger. Once determined, he directed his work for the development of techniques for RNA and DNA sequencing. Subsequently, he was able to publish the sequencing method for DNA through dideoxy sequencing, also termed as chain sequencing, a method in 1980. This method utilizes the replication enzyme, DNA polymerase which helps in the synthesis of a new DNA strand. Hence, the sequence of the strand may be revealed through the formation of the complementary strand of DNA. It occurs by linkage of 3' OH of the free nucleotide to the 5'-hydroxyl end of a primer. A primer is a small single-stranded nucleotide strand that can hybridize with a template DNA and subsequently extend by addition of nucleotides. The Sanger method required the presence of dNTPs (nucleotide triphosphates), and a few 2'3'-dideoxynucleotide triphosphates (ddNTPs). ddNTPs are nucleotides that lack the reactive 3'-OH and therefore, once incorporated into the growing DNA strand, DNA synthesis gets terminated (Figure 3.2).

During the process, a mixture consisting of dATP, dTTP, dCTP, dGTP, and one specific ddNTP is used. Extension of the primers continues until a complementary ddNTP attaches to it and thus terminates the process. As result, varying lengths of DNA strands are obtained each having a ddNTP at their ends. All the newly synthesized DNA fragments are detected by radiography due to the incorporation of radioisotopes of phosphorus and sulfur in the dNTPs. The obtained fragments of DNA can be separated by polyacrylamide gel electrophoresis on the basis of size. By running separate reactions with all four types of ddNTPs, the sequence of the template strand can be obtained.

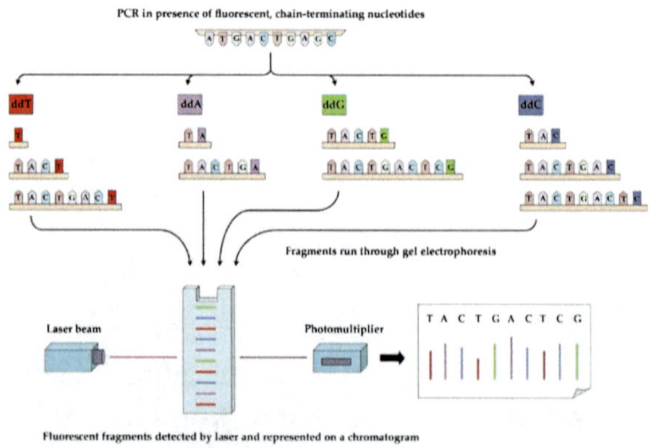

Figure 3.2: Methodology for Sanger sequencing. (Source: *http://www.vce.bioninja.com.au/aos-3-heredity/molecular-biology-technique/sequencing.html*).

3.2.2. Next-Generation Sequencing

Following the Sanger sequencing method, massive advances in sequencing methods have taken place. It is now possible to sequence millions of reads in a single reaction through commercially used technologies, the next generation sequencing (NGS). This technology has been applied to reveal important information about fungal genomes. Resequencing of *S. Cerevisiae*, denovo sequencing of the xylose fermenter, Pichia stipites, transcriptome analysis and fungal characterization have all been achieved through exploitation of this method.

3.2.2.1. Pyrosequencing by Roche/454 GS FLX Pyrosequencer

In 2004, the first commercial NGS was produced by Roche Diagnostics (formerly known as 454 Life Sciences. Pyrosequencing method that it uses involves the action of a light emitting enzyme, luciferase. It results in the emission of light when each nucleotide is incorporated by DNA polymerase. Due to development of latest instruments, sequencing kits, and reagents it has now become possible to sequence up to 1 million reads in a single 10 h run.

This sequencer works in three basic stages; preparation of a library of single-stranded template DNA, followed by an emulsion based method for clonal amplification of the library and then finally sequencing by synthesis.

For the preparation of DNA library, sample DNA is broken down into small fragments of single-stranded DNA (300–800 base pairs). To each fragment, 3' and 5' end specific universal adapters are attached. A universal adapter constitutes of 44 bases (20-base PCR primer, a 20-base sequencing primer, and 4-base initiating sequence.

In the next stage, the prepared DNA library is allowed to mix with DNA capture beads (up to 25 μm in size). These beads are made up of one of the adapter primers and therefore it binds with a fragment from DNA library. To ensure that single capture bead ligates with a single fragment of DNA library, the ratios of both mixtures are chosen carefully. Using amplification reagents, the complexes formed are allowed to emulsify. As a result, microreactors are formed which consist of one capture bead bound to one unique sample-library fragment. Simultaneously, the library fragments are allowed to amplify via thermal cycling. Several million copies of unique library DNA per bead are obtained at the end. The emulsion is then broken down by enriching the DNA-positive beads and depositing them to a PicoTiterPlate (PTP). These beads are covered with packing and luciferase and sulfurylase enzyme beads.

Finally, the sequencing is carried out by the process of synthesis. Sequential and ordered flow of nucleotides across the PTP takes place. DNA polymerase attaches complementary NTPs on the template strand helping in formation of a double strand. The incorporation of different nucleotides takes place at different rates, therefore, extension of strand also occurs at different rates. The sulfurylase enzyme present in the beads converts the generated free pyrophosphate into ATP. Luciferase enzymes catalyze the oxidation of luciferin using this ATP in the presence of light. Number of bases incorporated is directly proportional to the light intensity and therefore a count of the number of photons emitted per nucleotide (using a charge-coupled-device camera) indicates the DNA template sequence.

3.2.2.2. IGA (Illumina Genome Analyzer)

This sequencer has been in use since the year 2006. It allows sequencing of up to 300 gigabases in a single run. It can sequence fragments that have shorter read lengths than those for 454 sequencing (150 bp). It is the most cost-effective method for production of a vast amount of data. This sequencer also works on three basic steps; DNA library preparation, generation of clonal clusters and sequencing by synthesis (Figure 3.3).

Preparation of a DNA library is done by sonicating (using ultrasound) or nebulizing(through vaporizing) the DNA sample to generate fragments. During this stage, ligation of sequencing adapters is also done.

This stage is followed by generation of clonal clusters. This is carried out by immobilizing DNA templates to be sequenced onto a flow cell made up of silica, containing eight lanes. This allows sequencing of eight individual DNA templates per slide as each lane can be loaded with a separate DNA library. The lanes are covalently attached to forward/reverse 'lawn' primers. The free 3' ends of adapter-ligated template DNA gets captured by these primers in the flow cell resulting in formation of U-shaped bridges. Solid-phase bridge amplification occurs through addition of unlabeled nucleotides. As a result, clonal clusters that are double-stranded are formed. Once formed, the reverse strands are removed from the newly synthesized double strand of DNA. Subsequently, sequencing primers are allowed to hybridize on its 3' ends, ensuring that sequencing of clusters occurs from the same end. Following this step, the flow cell is moved to IGA for sequencing.

Sequencing by synthesis in IGA is carried out by fluorescent terminator nucleotide incorporation. To each of the growing DNA chains, DNA polymerase adds a terminator nucleotide (dNTP). Once the cycle is completed, IGA takes an image of the fluorescent dye allowing identification of the added nucleotide base. After identification, the terminator dye is cleaved and the next nucleotide is allowed to attach to it. The lengths of the sequenced clusters are all same as they are governed by the number of cycles undertaken.

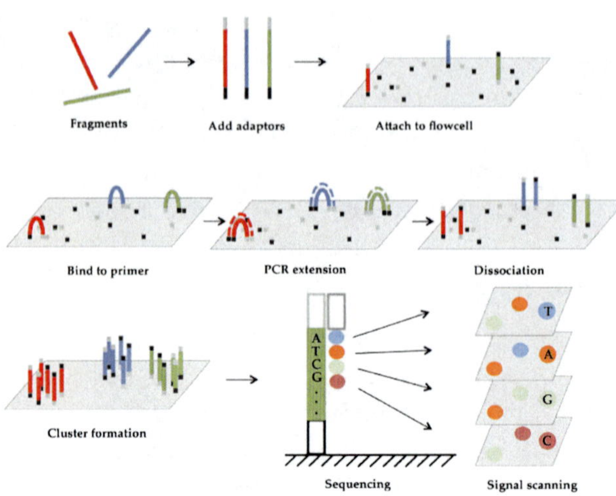

Figure 3.3: Methodology used for Illumina Genome Analyzer. (Source: *Yuan Lu, Yingjia Shen, Wesley Warren and Ronald Walter (2016). Next Generation Sequencing in Aquatic Models, Next Generation Sequencing – Advances, Applications, and Challenges, Dr. Jerzy Kulski (Ed.), InTech, DOI: 10.5772/61657*).

3.3. TOOLS USED IN BIOINFORMATICS

3.3.1. Location of Gene Homologs

An essential part of genomic studies is searches for sequence similarities. It enables scientists to predict functions of unannotated genes and to identify homologous sequences and conserved structural motifs. The genetic data published on databases like NCBI has enormously increased during the recent decade. To exploit this data, techniques for search sequence similarities require being efficient. Some useful online resources for fungal genetic databases are listed in the Table 3.1.

Table 3.1: Useful Resources for Online Databases Fungal Genes

Database	URL address
FGI	http://www.broadinstitute.org/annotation/fungi/fgi/index.html
SGD	http://www.yeastgenome.org/
CGD	http://www.candidagenome.org/
AspGD	http://www.aspgd.org/
CADRE	http://www.cadre-genomes.org.uk/
Aspergillus fumigatus database	http://old.genedb.org/genedb/asp/
CandidaDB	http://genodb.pasteur.fr/cgi-bin/WebObjects/CandidaDB
NCBI	http://www.ncbi.nlm.nih.gov/
Fungal Genomes Central	http://www.ncbi.nlm.nih.gov/projects/genome/guide/fungi/
Wellcome Trust Sanger Institute	http://www.sanger.ac.uk
EMBL	http://www.embl.de/
DDBJ	www.ddbj.nig.ac.jp/
SWISSPROT	http://expasy.org/sprot/
Fungal Tree of Life	http://aftol.org/
CGOB	http://cgob.ucd.ie/

(Source: *Whiteway, M., & Bachewich, C. (2005). Fungal Genetics Fungi (pp. 35–63): John Wiley & Sons, Ltd.*)

3.3.1.1. Local and Global Alignment

Homology among genes can be inferred using two main methods; global and local alignments. When two sequences are aligned over their entire lengths, it is called a global alignment. This is especially useful when the lengths and sequences of the aligned sequence are similar. On the other hand, alignment of two sequences is carried out only at regions that are highly similar in a local alignment. It is particularly informative as it allows identification of conserved domains which may have important functions.

Global alignment of sequences was first done using dynamic programming by Needleman and Wunsch in 1970. This is a computational technique which allows determination of highest scoring alignment between two sequences. This method was later adapted for local alignment. This technique works on the basis of a scoring matrix where each position in a row and column corresponds to the aligned bases or amino acids in the sequence. In the matrix, the first row and column are filled with zeroes, followed by iterative filling of the rest of the cells with scored depending on values in the neighboring cell. Correspondence of an identical base or amino acid is represented in the matrix cells by addition of a match score to the neighboring diagonal square. Alternatively, the maximum score is determined from cells above by adding a gap penalty, which is a negative number. In local alignments, scoring is started from the highest scoring cell in the matrix, following a trace path eventually to a cell that scores a zero.

3.3.1.2. BLAST

BLAST, basic local alignment search tool (is the most prevalently used tool for identifying homologs within a sequence database. It is sensitive and efficient searching tool that works on both nucleotide and protein sequences.

The algorithm for BLAST initiates when search is "seeded" with a query word consisting of letter from the query sequence. The query words along with related words (with slight variations in sequences) are located. These words are then scored using a scoring matrix and thus produces "neighborhood." The words that are closely related to the query word are determined using a neighborhood threshold "T." The greater the value for T, the higher will be the closeness of the aligned sequences.

BLAST algorithm begins by aligning the original query word above a certain value for T. The alignments are then extended in both directions and the alignments are simultaneously scored for matches, mismatches, and gaps. Determination of maximal length of the alignment is done by considering both the number of aligned positions and the cumulative score of the alignment. The extension of alignment is stopped when the number of mismatches exceeds a certain value (X). The result of this alignment is termed as the high-scoring segment pair (HSP). The algorithm also defines a score threshold for the HSP that are to be recorded in the result file for BLAST.

The last step involves determination of biological significance of the HSP. E-value is assigned to each alignment as it permits counting the number of HSPs that have a score greater than S by chance alone. Conclusively, the HSP is biologically significant if the score is above S by chance alone. A small E-value indicated greater biological significance.

3.3.1.3. FASTA

Similar to BLAST, FASTA is computational programs used for rapidly align pairs of proteins or DNA sequences. It was the first tool used for similarity searching in database.

The algorithm for this program starts by finding subsequences above a particular word length from the database sequence to subsequences of the query sequence. Ktup is the term used for the parameter of word length which is equivalent to *W* used in BLAST algorithm. FASTA works by creating diagonal lines for matching residues on a dot plot. Next, location of diagonal regions which contain maximum ktup matches with minimal distances between them is carried out. The highest scored regions are displayed as local alignments which do not have gaps. The adjacent diagonals are subsequently joined together in order to increase the alignment length. A joining penalty is invoked for every diagonal that is connected. Score of the alignment is assigned by adding net scores of diagonal regions excluding the penalties for joining them. The enlarged diagonal is given a score termed as init*n*. All of the highest scoring enlarged diagonals are sorted and arranged according to their scores using local alignment. The aligned sequences are then assessed for their significance in the final step. This is done by random generation of DNA sequences with similar sizes and compositions as the query sequence. The alignment is then scored for the probability that it happened by random chance (Figure 3.4).

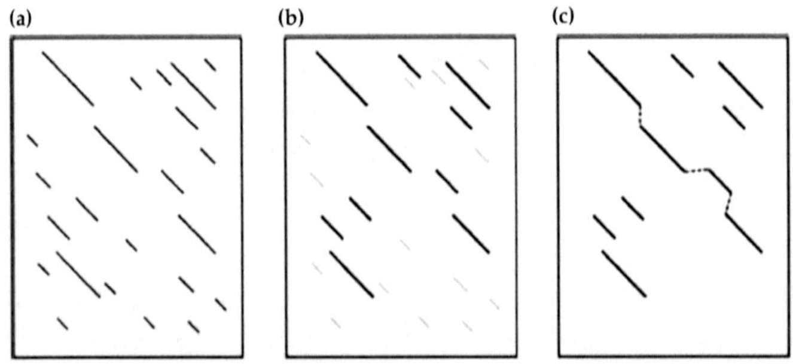

Figure 3.4: Sequence of the algorithm used for FASTA searches. A) Location of common words between the target and the query sequence. Close words are diagonally connected by lines. B) Diagonals that are on the top 10 are selected for the next step. C) Alignment of diagonals is done using a local alignment strategy. (Source: *Curran, B., & Bugeja, V. (2005). Fungal Genetics: A Post-Genomic Perspective Fungi (pp. 65–88): John Wiley & Sons, Ltd*).

3.3.2. MSA (Multiple Sequence Alignment)

Pairwise alignment is a useful tool for identifying homologs among annotated sequence and studying conserved regions among two sequences. To infer relationships between different families of DNA and proteins, we can use the multiple sequence alignment (MSA). Therefore, MSA is a more useful tool revealing more information. Moreover, this tool can be utilized for location of conserved residues and domains among hundreds to thousands of reported sequences which can uncover significant data about evolutionary relationships and physiochemical processes. MSA is employed extensively for phylogenetic studies and for primer designing in PCR reactions.

The algorithms used for multiple sequence alignment are much more complex than those used for pair-wise strategies used in BLAST and FASTA searches. In MSA, the most prevalently used algorithm is CLUSTAL which helps align the sequences of interest in a progressive manner. The steps involved in CLUSTAL are:

1. Pair-wise alignment of different sequences is done to assess how close their relationship is.
2. Scores of this alignment are represented in the form of a guide tree;

3. Progressive alignment of the sequences is carried out by the guidance of the phylogenetic tree. The sequences that are aligned first are the most closely related followed by alignment of additional sequence and group.

The progressive alignments are refined in CLUSTAL by using various alignment penalties. For instance, penalties for gap and extension are implemented in case of hydrophilic region. This is because it is more likely to contain regions of loops and random coils which commonly have gaps. Another example is a residue-specific penalty for glycine-rich region which is more likely to have gaps in the adjacent regions compared to other amino-acid rich regions (e.g., valine).

3.3.3. Gene Ontology

Genes that are derived from common and are likely to have the same functions are called as orthologues. This means that if the biological role of a gene is defined in one organism, the function of its orthologue in another species can be predicted. However, it is a difficult task to organize biological data from various databases, especially because each database uses variable terminologies for the same process.

This difficulty has been overcome by development of Gene Ontology (GO) Consortium. It has a vocabulary, for describing functions of genes, which is organized, common, and clearly defined. Ontology provides a platform for searching, storing, and querying terms within various databases. The GO Consortium contains details about annotations for 12 significant model organisms, including two fungal species, *Saccharomyces cerevisiae,* and *Schizosaccharomyces pombe*. The terminologies that are used in GO come under any of the three categories; cellular component, biological process and molecular function. These 12 annotated species are referred to as GO reference genomes. By comparing to the GO reference genomes, the annotation of fungal species can be greatly accelerated. Based on similarities to the sequences in model organisms, putative functions can be assigned to the *de novo* genes. The GO resource is very applicable for studying *de novo* fungal genomes as two fungal model organisms are reported in this database.

3.4. COMPARATIVE GENOMICS

3.4.1. Disease-Linked Families of Genes

It has been found that some families of genes are significant in causing virulence in species of fungi through comparative genomic studies. For instance, a comparative genomic analysis carried out using genomes from 34 fungi revealed the identity of several pathogenic fungal families, including *Botryotinia cinerea*, *Ashbya gossypii*, *Magnaporthe grisea*, *Sclerotinia sclerotiorum*, *Stagonospora nodorum*, *Ustilago maydis* and *Fusariumgraminearum.* During the evolution of phytopathogens, these fungal families duplicated with time into multiple species. The study also found the secretory proteins that were produced by the phytopathogens and categorized the gene families that consist of a secretome. The identified protein families were found to be associated with pathogenic processes. (e.g., degradation of cell wall). The genome sequenced for *U. maydis,* the corn smut fungus, revealed a vast secretory protein set that clustered together which constituted a total of around 20% of the total secretory proteins. The importance of these extracellular proteins was demonstrated by serious effects on virulence upon deletion of individual clusters.

Several gene families are found abundantly in pathogenic *Candida* species of fungi (*C. albicans*, *C. parapsilosis,* and *C. tropicalis*) relative to the nonpathogenic species. For example, comparative genomic analysis has revealed that Hyf/Iff proteins are found in all species of Candida but they are particularly abundant in pathogenic Candida species (11, 17, and 18 copies respectively). This protein family codes for cell wall components and is found to be rapidly evolving on the basis of sequence similarity. Because rapid evolution of proteins of cell wall is a frequently used mechanism for escape by pathogenic microbes, most probably, these proteins are involved in recognition of host. Agglutinin-like sequence (Als) proteins are also enriched among the pathogenic *Candida* species. These genes are well studied in *C.albicans*. They are known to play a role in formation of biofilm, adhesion to plast surfaces and to host cell surfaces. In a hospital setting, the ability of *Candida* species to adhere to plastic surfaces is particularly problematic. This is because these pathogens infect the patient's blood through medical devices. When the *Candida* species grow to form biofilms, a reduced susceptibility to antifungal drugs is observed.

3.4.2. Synteny

When two or more loci are present on the same chromosome, the term synteny is used by geneticists. However, in postgenomic studies, this term is used to refer the relative order of evolutionarily close genes on the chromosome. Syntenic regions in two genomes, therefore, consist of a conserved order of multiple consecutive genes (figure 3.5). Breakage of synteny between two species may occur as a result of rearrangements in genomes of either or both of the species after evolving from a common ancestor.

Through comparative genomic studies of fungi, it has been observed that the conservation of syntenic structure is observed among closely related species and this conservation reduces as the relationship between the fungal species becomes more distant. For example, minimum conserved synteny is observed among the various subdivisions of Saccharomycotina family of yeasts. However, when members of the same subdivision are analyzed, they consist of large conserved syntenic regions. Using the *Candida* genome order database, it has been revealed that nine of the Candida genomes have numerous conserved syntenic blocks. Among the closely related species, a conserved order of genes is correlated to the phylogenetic analyses. The *Saccharomyces 'sensu stricto'* species consist of homologous chromosomes are almost collinear. They differ by only a small number of translocations and huge inversions. Extensive syntenic regions have been detected among *Coprinopsis cinerea* and *Laccaria bicolor,* two species in Basidiomycete phylum. Syntenic blocks are larger in regions that have low probability of meiotic recombinations and have no transposable elements (Figure 3.5).

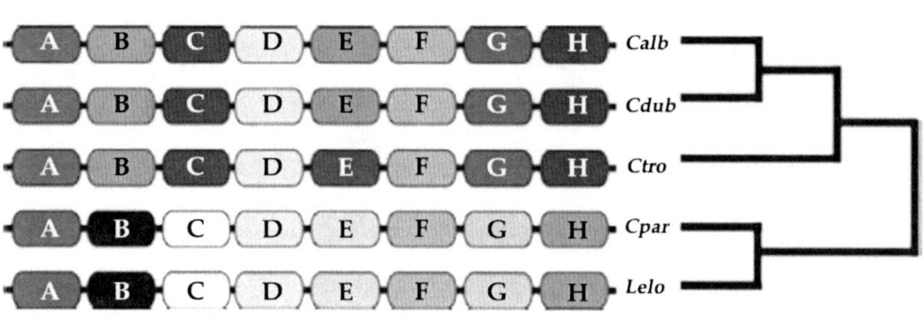

Figure 3.5: Synteny among five *Candida* species for of eight orthologous genes. Syntenic genes in all five species are A, Dand F; (represented by conservation of color between them. In *C. albicans*, *C. dubliniensis* and *C. tropicalis*, genes B, C, G, and H are syntenic. Genes B, C, G, and H are orthologous and syntenic in *C. parapsilosis* and *L. elongisporus*, even though their locations are

different compared to the first three species. (Source: *Curran, B., & Bugeja, V. (2005). Fungal Genetics: A Post-Genomic Perspective Fungi (pp. 65–88): John Wiley & Sons, Ltd.).*

3.4.3. Analysis of Metabolism in Silico

Because of the presence of a large amount of fungal genomic data, it is easy to make an *in silico* comparison of the metabolic repertoire in different species. Comparison of metabolic pathways in one species makes it possible to find differences and similarities in other species. Such a comparative metabolic analysis among pathogenic and nonpathogenic species of fungi reveals useful information about potential disease mechanisms. Additionally, it allows the investigation of metabolic pathways that enable a species to degrade a specific substrate with another species that is unable to do so.

It has also been revealed through comparative genomic analysis that some genes in fungi are not located randomly. A study done on *S. Cerevisiae* genome has shown that the genes involved in a particular metabolic pathway are clustered together within the genome. Similarly, in filamentous fungi including *Aspergilli* species, genes involved in secondary metabolism are clustered together in their genomes.

Studies on syntenic conservations done on nine *Candida* species have shown that among all the analyzed metabolic pathways, 20% depict at least some evidence of clustering. One of the metabolic pathways that were clustered together is involved in metabolizing *N*-acetylglucosamine (Nag) to fructose-6-phosphate. Initially, it was thought utilization of Nag as alternative source of carbon is a significant factor for virulence in pathogenic *Candida* species. There are three genes that help in conversion of Nag to fructose-6 phosphate including hexokinase kinase (HXK1), Nag-6-phosphate deaminase (NAG1) and Nag-6-phosphate deacetylase (DAC1). In *C. albicans*, these genes are present in the form of cluster called 'Nag regulon' and these enzymes act sequentially on the substrate. The Nag regulon is shown through synteny analysis to be highly conserved among all species of *Candida*. The Nag regulon is conserved in both pathogenic (*C. albicans*, *C. tropicalis,* and *C. parapsilosis*) and nonpathogenic fungal species (*C. dubliniensis*, *Lodderomyces elongispors* and *Debaryomyces hansenii*). This indicates that utilization of Nag isn't a significant virulence factor. Even though Nag regulon is absent in *S. Cerevisiae*, this species can be induced to utilize Nag by introduction of NAG genes isolated from *C. albicans*.

3.4.4. Horizontal Gene Transfer

The exchange of genes between different strains or species is called as horizontal gene transfer (HGT). It involves introduction of new genes, which may be homologs of already present genes or may be part of families that are completely new, into the genome of the recipient. Genomic analysis of bacteria has shown that this process is significant in evolution, for obtaining drug resistance, and for catabolism of specific amino acids that act as virulence factors. Several approaches may be employed for detection of horizontally transferred genes within a genome. One method is by location of genes through an atypical codon or base usage pattern. Alternatively, this can be achieved by sequence similarity searches using query gene against the genes in a database. This helps in finding the top scored matches within the database. Even though these methods are quick and automated, their degree of accuracy is not very high. A drawback in using this approach for studying HGT was seen when the initial reports of the human genome showed 223 genes that came from the bacterial pathogens. This report was based on the results from BLAST search. However, phylogenetic studies revealed that were not recently introduced into the human genome from bacteria via HGT. Phylogenetic inference has proved to be the most reliable method for detecting HGT. The topological disharmony among trees belongs to two different gene families is often explained by the concept of HGT.

In fungi, the gene transfer process is insignificant. However, in fungal genomic evolution, the significance of HGT has been proved through the comparative analysis of the vast fungal genomic data available.

For example, since *S. Cerevisiae* diverged from the closely related *A. gossypii,* it has incorporated around 13 bacterial genes through HGT. Even though they constitute of only 1% of the *S. Cerevisiae* genome, these 13 genes have major contributions towards functional innovations. These functions include the yeast's ability to grow under anaerobic conditions, synthesis of biotin and utilization of sulfate from some organic sources. Other examples for HGT in fungi are; obtaining of bacterial metabolic genes by *C. parapsilosis* and obtaining of a toxin gene (ToxA) by *Pyrenophora triticirepentis* from *Stagonospora nodorum* resulting in *Pyrenophora* infestations of wheat.

The mechanisms of gene transfer in fungi are not yet well-understood. Moreover, mechanism for uptake of DNA has not been detected in this organism. However, a conjugation between yeast and bacteria has been seen. The horizontal gene transfer most likely occurs due to the fact that fungi live in close proximity to other organisms as they are saprobes.

3.5. GENOMIC STUDIES AND TREE OF LIFE OF FUNGI KINGDOM

3.5.1. Phylogenetics

Phylogenetic studies enable prediction of evolutionary relationship among sets of populations, species, individuals or genes. It is done by arranging them in a logical manner which helps infer this relationship. A representation of evolutionary tree is used for inference of historical development of species. Fungal systematics using traditional methods can be employed for assigning species of fungi into specific genera and families. These methods are based on vegetative cell morphology, sexual status, physiological responses to fermentation and growth tests of fungi. Such approaches are especially desirable as the fungi do not show many morphological characters and hence lack any record in fossils. Among fungal species, sequence data from DNA, RNA, and proteins have been used to reveal evolutionary relationships. A distance matrix is often used to process aligned sequence. In such an approach, small distance among two species will represent a close relationship while a larger distance correlates to distantly related species. These distance matrices can be represented by phylogenetic algorithms including neighbor-joining, minimum evolution, and UPGMA.

To infer fungal trees of life, several phylogenetic markers may be used. Such a marker should ideally be present ubiquitously within the species of interest but as a single copy, should be less vulnerable to HGT and should consist of sites that are slow-evolving. 18S ribosomal DNA is, therefore, frequently used for determining relationships of fungi with other organisms. Analysis of single genes, however, depends on phylogenetic markers that have an evolutionary history, which in turn reflects that of the entire organism. This assumption of often violated. Additionally, single genes consist of limited amount of sites, and hence, have a limited resolution. Alternative to single-gene phylogenetics, multigene phylogenetics can also be used. It works by combining all phylogenetic markers. Two methods that are commonly used to achieve this are; concatenated multigene phylogeny reconstruction and supertree analysis.

3.5.1.1. *Concatenated Multigene Phylogenies*

This process results in production of a large super alignment through appendage of may aligned sequences. This combining of the data enhances the phylogenetic accuracy, usefulness, aids in resolving nodes and basal branching. Concatenation of universally distributed genes has helped develop several phylogenies of various species. This process has also been applied to redevelop evolutionary relationships of the fungal kingdom. This was done by employing Fungal Tree of Life consortium (Table 3.1) using the housekeeping genes from a total of 199 species of fungi; 18S rRNA, 28S rRNA, 5.8S rRNA, elongation factor 1-alpha and two RNA polymerase II subunits (RPB1 and RPB2). It helped demonstrate the evolutionary history of the phyla of fungi. Additionally, it revealed that during the early fungal divergence, the development of novel mechanism for spore dispersal took place simultaneously with the loss of spore flagella hence increasing terrestrial fungal diversity.

3.5.1.2. *Supertrees*

In this method, single genes help derive single input trees. All the input trees are put together to generate a single representation of phylogeny of species. (Figure 3.6). Comparative analysis of the fungal genomes showed that only 1% of the total genes in fungi are universally distributed. This means that when multigene phylogenies are redeveloped, 99% of the fungal genes are ignored. Supertree method enables usage of 100% of the genes which is ideal.

In this method, a set of input trees constructed from full or partial overlap of taxa are employed for generations of phylogeny. The set of input trees is therefore taken as an input and gives the output of a phylogenetic tree that is a representation of all the input trees. Maximum genes are used in this method and hence the developed phylogeny tree represents the entire genome. A total of 42 genomes of fungal species were used to identify 4805 families of genes which were used for construction of individual phylogeny trees. Supertree method helps summarize the complete set of data. Results from this study showed the presence of a single clade for Saccharomycotina, including *C. albicans* and its close relatives (figure 3.6). Similarly, *S. Cerevisiae* and close relatives, undergone duplication of whole-genome, are present as a monoclade. As more data on fungal genomes will become available, supertree techniques may be employed more usefully for development of phylogenetic trees of life for fungi.

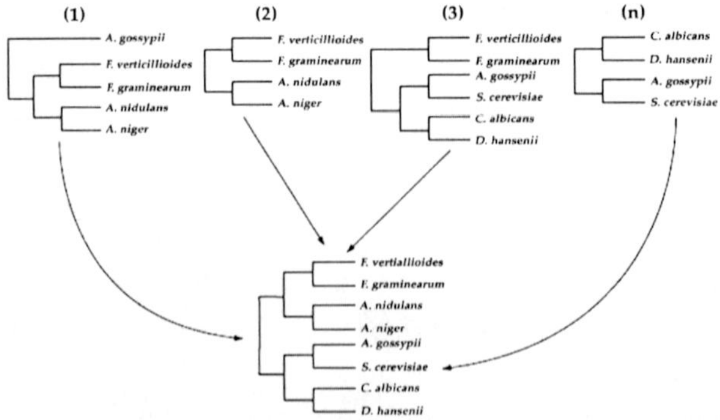

Figure 3.6: Representation of reconstruction of a supertree. It takes thousands of input trees to produce one Supertree for phylogenic studies. (Source: *Curran, B., & Bugeja, V. (2005). Fungal Genetics: A Post-Genomic Perspective Fungi (pp. 65–88): John Wiley & Sons, Ltd.*).

3.6. ONLINE FUNGAL GENOMIC SOURCES

3.6.1. National Centre for Biotechnology Information (NCBI)

NCBI (Table 3.1) is a very useful tool that consists of databases for published sequences, supports biological annotations, and tools for bioinformatic analysis.

It has a separate section for genome of fungi which is termed as Fungal Genomes Central (FGC). It links to 309 sequencing projects (both ongoing and completed) for fungal genomes. It also has links to raw sequence reads produced by a few sequencing centers for fungi (Table 3.1). Additionally, mitochondrial DNA sequences from 80 fungal species are also stored here. This vast fungal genomic database can be searched against a single gene of interest using fungal BLAST search. Automated detection system for homologous genes of fungi is available at NCBI. The current list of genes is composed of 9000 fungal genes. This tool also has a search engine for fungal taxonomy consisting of names and phylogenetic lineages of the NCBI-recorded species among the fungal kingdom.

3.6.1.1. Databases for Genomes of Saccharomyces, Candida, and Aspergillus

A specific database for *Saccharomyces,* Saccharomyces Genome Database (SGD) (Table 3.1), was published in 1997. It is a popularly used database that is present at the Stanford Human Genome Center. It allows the users to view complete genome of *S. Cerevisiae* including its genes, their products, phenotypes of the mutants and the literature regarding these data. SGD does not consist of only primary sequences of genes, in fact, it obtains information about DNA and protein sequences from primary providers, such as, GenBank, EMBL, DDBJ, and SwissProt (Table 3.1). The gathered data is then assembled into data sets which are useful for users. Therefore, SGD is a merged form of several primary databases which eliminates the need for searching resources separately.

This highly annotated database along with its records for literature is carefully secured and assisted by SGD curators. Each of the *S. Cerevisiae*'s genes is assigned to an appropriate publication extracted from PubMed. SGD allows users to use a refined text-based search engine to look for information on genes' names, genes, proteins, and authors' names. Additionally, SGD enables BLAST searches, allows access to metabolic pathways of yeasts, literature related to yeast, information about gene expression obtained from several microarray studies, and study relative positions of genes on the chromosomes.

On each locus page in SGD, information about the genes, their names and their systematic names are located. Based on experimental evidence, SGD can also provide information about the genes' features. They can be 'verified' or 'uncharacterized' depending on the evidence from the literature. The 'description' of the genes displays significant amount of information about both the genes and their products. Each associated product is given a term according to gene ontology which is based on its molecular functions, its cellular locations, and its putative role in a biological process. Moreover, a section to view mutant phenotypes is also present on this page which lists various types of mutations and their corresponding phenotypes. Within this locus page, links to information about sequence and the literature supporting it are also located.

Candida Genome Database (CGD) became available to users online in 2005. It provides extensive information about *Candida* species regarding its genetics and pathogenesis. Prior to its launch, information about the

Candida genome and their products was available on three different websites. Candida genome was sequenced and distributed by The Stanford Genome Technology Center (Table 3.1). Both *Candida*DB and *Candida* Working Annotation group (Table 3.1) consist of annotated genes of *C. albicans*. Information was pooled from all these websites and the data has been expanding since them. CGD and SGD work on the same framework. Both have the same softwares, user interfaces, and data structures.

In 2009, the *Aspergillus* Genome Database (AspGD, Table 3.1) was published. It provides scientists with information about genetics and molecular biology based on *Aspergillus* genome. Information about *Aspergillus* genome is currently available of numerous databases such as the Central *Aspergillus* Data Repository (CADRE) (Table 3.1), which has information which is oriented to clinics and patients, The Broad Institute (Table 3.1) contains the genome for *Aspergillus*, database for the *Aspergillus fumigatus* (Table 3.1) and several other websites that aim towards projects for different species of the *Aspergillus.* These resources from these individual databases are linked and complemented in AspGD where an in-depth curation of the scientific literature regarding the data is done. The framework for AspGD is similar to CGD and SGD as previously described. Currently, AspGD focuses on curation of many species of *Aspergilli* including the best-characterized, *A. nidulans*, *A. fumigatus*, *A. flavus*, *A. oryzae*, *A. niger*, *A. clavatus*, *A. terreus* and *Neosartotya fischeri*.

3.6.2. FGI (The Fungal Genome Initiative)

The Broad Institute of Harvard and MIT have created a partnership for The Fungal Genome Initiative (FGI, Table 3.1). This was initiated by several devoted fungal geneticists and biologists due to their realization that the sequencing projects for the fungal genome were very slow which is a barrier to research in biomedicine and evolution. Instead of individual sequencing projects for fungal species, simultaneous sequencing of multiple organism would be carried out. The criteria for selection of fungal organisms was based on its importance in health and commercial activities of humans, its role in comparative genomic and fungal diversity studies, genetic resources and current researches related to the organism.

Three main aspects of diversity in fungi were initially selected, which included the following 15 fungi:

1. Medical; e.g., *Rhizopus oryzae*, cause of mucormycosis;
2. Evolutionary; e.g., *Neurospora discreta*, study of population genetics;
3. Commercial; e.g., *Aspergillus flavus*, source of aflatoxin in food.

Subsequently, 28 other fungal genomes have been focused. To employ comparative genomics, the strategic clusters that are chosen include; *Candida*, *Cryptococcus*, *Aspergillus*, *Fusarium*, *Histoplasma*, *Coccidioides*, *Penicillium*, *Neurospora*, *Schizosaccharomyces*, *Puccinia*, and *Schizosaccharomycetes*.

FGI has successfully sequenced 50 fungal genomes currently which are displayed on its homepage (Table 3.1). Information regarding several aspects of the genomes are included; genome maps, sizes, densities, and GC content. FGi has incorporated a BLAST search tool and a text-based search which facilitates its users to search for genes through their function or protein domains. Synteny maps can also be accessed by selecting evolutionarily close species present within a single cluster.

3.7. CONCLUSION

Several fungal species that are significant to humans in several aspects have been sequenced successfully. These include those that are pathogenic, such as, *C. albicans*, *A. fumigatus* and *Cryptococcus neoformans,* and industrially important species, *S. Cerevisiae,* and *Aspergillus niger*, which are necessary for brewing. Therefore, biases is observed in the fungal species that are being sequenced. It is expected that the number of sequenced genomes of fungi belonging to different phyla is going to elevate in the near future because of the small size of fungal genomes and reducing costs for sequencing. This data is expected to help us study pathogenicity and evolution, and reveal novel proteins that are medically and biotechnologically significant. Since this accumulation of data from various resources, the focus now is on the challenge to integrate molecular details in order to elucidate various cellular processes through systems biology which involves both experimental and theoretical approaches.

4
CHAPTER

POST-GENOMIC PERSPECTIVE OF FUNGAL GENETICS

CONTENTS

4.1. INTRODUCTION

There is an exponential increase in genomic data available in world databases. This deluge of information is simultaneously characterized and studied through the application of computer technology.

4.1.1. S. Cerevisiae: An Important Yeast Species for Research in the Post-Genomic Era

The first eukaryotic sequence genome recorded in an online database was that of yeast, *Saccharomyces cerevisiae*. The post-genomic era began when other eukaryotic genomes were sequenced simultaneously with developments in bioinformatics tools for storage, display, distribute, analyze, and compare the enormous amount of available data. For molecular analysis of eukaryotic cells, the genome sequence of *S. Cerevisiae* was the beginning point. This model organism has served to yield a wide range of molecular insights into eukaryotic organisms. This has been done by eliminating genes to observe phenotypes, profiling of global expression of mRNA and examining protein-protein interactions, and characterization of cellular metabolic products. This simple eukaryotic organism has, therefore, become a key to research in the post-genomic era.

4.1.2. Systems Biology

In pre-genomic research, sequential experiments are carried out which help in gene characterization. While, in post-genomic research is carried out by simultaneous analysis of biological data and subsequent recognition of patterns in datasets. The accumulation and analysis of increasing information about the DNA sequence became possible due to developed computational technologies. These involve automation of DNA preparation, large-scale DNA sequencing, storage, and processing of the generated data. Study of genomes is called genomics, therefore, global analysis of biological molecules is given the suffic –omic. For example, the study related to sequence and expression of mRNA is called as *transcriptomics*, while that of proteins is called as *proteomics*, which also involves the study of protein structure, levels, and interactions with DNA and proteins.

To characterize genomics, intensive technology for automatic and computational analysis of DNA sequences has been developed. For example, simultaneous identification of DNA or RNA sequences is carried out by DNA microarray technology, and multiple protein samples are identified using mass spectrometry. The development of these platforms has been greatly facilitated by *S. Cerevisiae*. Moreover, this microorganism has become an integral part of a newly developed biological research field, systems biology. This emerging field involves an approach merging biology, engineering, physics, mathematics, chemistry, and computer science to enable designing of models to represent molecular interactions in order to study biological phenomena.

4.2. GENOMICS

4.2.1. Analyzing Encoded Information

The study of genomes includes several aspects; comparative analysis between genome sequences revealing their relatedness and assigning putative molecular functions to genes based on their sequences. Several fungal genomes have been sequenced, including *S. Cerevisiae*, *Schizosaccharomyces pombe,* and *Neurospora crassa,* which serve as model organisms for research. Sequencing projects for more than 400 fungal genomes have currently been recorded at NCBI which includes both completed and on-going projects. These include a wide variety of fungal species that are agriculturally and commercially important and are significant to human health. Comparative analysis of these genomes enables us to improve our understanding about the evolutionary history of fungi. A lot of information revealed from this data will also serve to improve our knowledge about both animal and plant biology.

4.2.2. Recognition of Pattern amongst and within Sequences of the Genome

The computational analysis does not directly provide information about the function of a gene but it reveals useful data regarding topology and evolution of genomes. it uses various approaches like pattern recognition which involves identification of tracts of DNA that begin with for example the codon for methionine which runs for another 99 codons without hitting a nonsense one (Figure 3.6). These sequences are termed as open reading frames or ORFs and are categorized as highly probable to code for proteins (Figure 4.1). Some highly developed programs are even able to recognize sequences at the intron-exon boundary are therefore able to exclude the introns. An enormous amount of detail about genome can be elucidated including gene numbers, sizes, sequences, spacing within a chromosome, the presence of introns and much more. Around 6000 genes in *S. Cerevisiae* genome have been identified by employing such programs. It has also been that compactness of yeast genome is very high and 70% of it encodes for proteins with very few intronic regions. Moreover, regions producing non-coding RNA, like rRNA, dnRNA, and tRNA have been located and identified. Analysis of pattern-recognition in S. pombe has shown that its genome has 600 lesser genes coding for proteins compared to *N. crassa*, which is its close filamentous course and consists of around 10,000 genes. In both *N. crassa* and *S. pombe*, intronic regions are present, unlike *S. Cerevisiae*. They also have a lower percentage of genome coding for proteins and also lower gene density compared to *S. Cerevisiae* (Table 4.1).

```
      GTC
     TGT
    ATG
5 - ATGTCGAATTCGCCTATAG - 3
3 - TACAGCTTAAGCGGATATC - 5
                     ATC
                    TAT
                   ATA
```

Figure 4.1: Searching for an ORF. Six different reading frames are searched while looking for ATG codon for methionine. Before a non-sense codon is hit, the search is continued for a further 99 codons. (Source: *Curran, B., & Bugeja, V. (2005). Fungal Genetics: A Post-Genomic Perspective Fungi (pp. 65–88): John Wiley & Sons, Ltd.*).

The identified ORFs of a genome can be compared to all other ORFs inorder to reveal evolutionary relationships between them. Through this analysis, it has been found that *S. Cerevisiae* evolved from an ancient organism through gene duplication as 55 parts of its chromosomes contain three or more homologous gene pairs, despite it being a haploid organism. This helped hypothesize a model where two diploid yeast ancestors, consisting of 5000 genes each, fused together to form a tetraploid. As this species evolved, the majority of the duplicated genes were eventually lost by deletion making it a haploid organism with around 5500 genes. The duplicated regions of the yeast encodes for around 13% of its proteome. 2.4% of the yeast genome was found to be composed of transposable elements. Large-scale genomic duplication was not detected in *S. pombe* through intragenomic searches even though few evidences supported the fact that this species evolved through gene duplication. 10% ORFs of its genome were defined to be paralogous (Gene duplication within the same species). Comparison of ORFs helped identify 11 intact transposable elements which account for 0.35% of the genome (Table 4.2).

Contrary to both of these species, ORFs analysis in *N. crassa* failed to support gene duplication, large-scale genome duplication, and transposon activity during its evolution. Less than 4% of the genome of *N. crassa* is redundant, consisting of paralogous ORFs suggesting that no genome duplication, rare gene duplication, and very limited transposon activity occurred in the history of species (Table 4.2). A repeat-induced point mutation (RIP) can explain the rareness of paralogues and transposable elements in *N. crassa*. This process occurs during the sexual development of fungi, when up to 1 kb lengths of duplicated DNA sequences undergo hypermutations.

4.2.3. Functional Genomics in Fungi

The process of assigning biological functions to genomic sequences is called as functional genomics. It has many aspects including *in silico* analysis for attribution of potential functions to few ORFs based on sequence similarities to sequences whose functions have already been reported in databases. Another alternative is to exploid reverse genetics, observing phenotype by mutating the gene of interest with unknown function. If this fails, comparative genomics can be implied to assign possible functions to a gene. This is done through large-scale genomic comparisons by virtue of the fact that some sequences absent in specific genomes but are present in others (identification by association).

4.2.4. Studies Involving In Silico Analysis

Annotation of RNA, DNA, and protein sequences into databases was initiated long before the advent of genome-sequencing projects. Exponential increases in this data urged the need for organizing and improving the accessibility of this data. Annotated DNA sequences and function of their encoded proteins together reveal invaluable information about putative or confirmed protein functions of ORFs that are newly identified. Annotation of around 50% of *S. Cerevisiae* genome was done through BLAST searches of these resources. This information was then used to uncover the percentage of genes that are dedicated for several biological processes. It revealed that more than 600 proteins are part of its metabolism, more than 400 proteins are part of intracellular trafficking or protein targeting and around 200 are transcription factors. Presence of certain genes that was previously not confirmed in *S. Cerevisiae* was identified through this method. For example, the histone protein H1 was located on chromosome XVI and a γ-tubulin gene, which remained undetected previously despite extensive effort, was located on chromosome XII.

4.2.5. Studies Involving Reverse Genetics

The limitation of *in silico* analysis is that it can only annotate newly discovered ORF sequences employing previously-identified functions. Usually, the annotation is based on protein domains that are present and therefore, the functions assigned are general (like protein kinase or transcription factor). The specificity of this assignment can be improved through phenotypic analysis when the gene of interest is deleted or disrupted from the organism's genome. It was due to this reason that scientists exploited *S. Cerevisiae*'s ability to carry out homologous recombination in order to generate mutant strains with specific deletions of ORFs. It was expected that biological function of the specific deleted ORF may be revealed through this process. 758 ORFs with unknown functions were sequentially deleted and the resultant mutant phenotypes observed in the EUOFAN project. This was followed by a more focused range of deletion mutants carried out by several other European laboratories. The most extensive and challenging work was eventually done a transatlantic consortium. They generated hybrid DNA molecules consisting of selectable markers and 5' and 3' flanking region of the ORF of interest through PCR. They carried out the recombination of these molecules by transforming them into wild-type strains and then selecting the positive recombinants. Successively, all 6000 genes were deleted in this way. Transposons have been used to induce insertional mutagenesis In order to develop several mutant yeast libraries. Subsequently, phenotypic characterization of several of these strains has been done. Since 1996, when the 3000 ORFs of yeast genome with unknown function were discovered, the biological roles of more than 1000 of them have been identified till date.

Molecular approaches to achieve gene knockout have also been invaluable for studying many fungal species. The pathogenicity of several fungal species including *Candida albicans*, *Candida dubliniensis*, *Cryptococcus neoformans*, *Aspergillus fumigatus*, *Histoplasma capsulatum* and *Coccidioides immitis* has been revealed through gene knockout by the use of appropriate selectable markers and procedures for transformation. This basic approach can be useful in investigating *S. Cerevisiae* which will help develop antimycotic drugs in fungal post-genomic research. Homologous recombination frequently occurs in these pathogenic species and their genomes have already been completely or partially sequenced. Because homologous recombination is not possible in all pathogens, the approach for gene knockouts to elucidate pathogenicity is not as simple. The virulence varies among different isolates, different conditions. Diploid Candida species requires two independent events of deletion to create knockouts. Reverse genetics is nevertheless being applied to numerous fungal species. However, not all fungal species can be studied using this strategy. alternative methods, such as, 'guilt by association are being developed to assign the ORFs of unknown functions with biological functions.

4.2.6. Studies Involving Association

The second fungal genome to be completed was that of *S. pombe*, which was also the sixth annotated eukaryotic genome. This success made it possible to compare two yeast genomes, *S. Cerevisiae*, and S. pombe for the first time to one another and to other organisms. Comparative analysis with the simple multicellular eukaryotic *Caenorhabditis elegans* revealed that 14%of the total ORFs were unique *in S. pombe,* 16% were homologous *to S. Cerevisiae* and around 66^ were homologous in both *S. Cerevisiae* and *C. elegans.*

Neurospora is a fungal species that is multicellular and has 28 types of cells that differ in morphologies. Interestingly, it has over 11000 non-paralogous genes. When it was compared to its cousin single-celled yeasts, *S. Cerevisiae* and *S. pombe,* expectedly, it showed very little homologs. Compared to other yeasts, this filamentous fungus has a more diverse range of environmental sensing pathways, transcription factors, sugar transporters, and metabolites. Moreover, cytochrome P450 domain-containing proteins are 38 in number, in contrast to as few as five such proteins in *S. Cerevisiae and S. pombe.* Further identification of groups of genes associated with specific fungal characteristics will be possible as more fungal genomes will become available. Information about species-specific genes and genes that are multicellular or yeast specific were revealed through comparisons of distantly related fungi. These predictions would become more accurate as more closely related genomes will be available. To date, much has been revealed about fungal biology through comparative analysis. Up to 200 putative proteins in *Neurospora* are similar to disease-associated genes in humans. Several of these have homologs in both *S. Cerevisiae* and S. pombe, 23 of these being related to cancer. These reports indicate the potential of fungi as model organisms for studying human disease pathways such as neurodegeneration. Information other than that revealed from the genome is required to study the development of fungal organisms. This is carried out by characterizing global changes in molecules of cells which follow the central dogma.

4.3. TRANSCRIPTOMICS AND PROTEOMICS

Initially, biologists explored DNA expression through mRNA transcript analysis by employing northern blotting. Similarly, 1-D and 2-D PAGE (Polyacrylamide gel electrophoresis) were used to monitor altered patterns of protein expression. Information revealed to scientists through these techniques was limited in terms of time and space. With the advent of the post-genomic era, it is now possible to exploit the sequenced genomes to study parallel alterations in profiles of expression of every gene through array technology. Detailed insights into locations and interactions of proteins, both characterized and uncharacterized are now being revealed. Rapid developments in this era allow extensive molecular analysis of information-rich molecules.

Table 4.1: Genome Topology of Model Fungi

Fungal Species	Organismal complexity	Genome size (Mb)	Protein encoding sequences	Gene density (excluding introns)
S.cerevisiae	Budding yeast	12.07	5500	1 every 2.09kB
S. pombe	Fission yeast	13.8	4900	1 every 2.53kB
N.crassa	Filamentous fungus	41	12000	1 every 3.1kB

(Source: *Curran, B., & Bugeja, V. (2005). Fungal Genetics: A Post-Genomic Perspective Fungi (pp. 65–88): John Wiley & Sons, Ltd.*)

Table 4.2: Evolutionary History of Model Fungal Genomes

Fungal Species	Evidence of former genome duplication	Evidence of gene duplication	Evidence of transposon activity
S.cerevisiae	Yes	Yes	Abundant
S.pombe	No	Yes	Abundant
N.crassa	no	no	Very little

(Source: *Curran, B., & Bugeja, V. (2005). Fungal Genetics: A Post-Genomic Perspective Fungi (pp. 65–88): John Wiley & Sons, Ltd.*)

4.3.1. Transcriptomics

Transcriptomics involves studying RNA sequences and their expression levels. The post-genomic era demands advances in the analysis of mRNA expression. Northern blots were able to allow comparison of expression of only one or two genes. Gene expression patterns need to be analyzed at a more global perspective. The first analysis on a fungus was carried out using *S. Cerevisiae*.

For the first time, PCR was used for amplification of 6400 ORFS of the yeast genome in 1997. Using a robotic printing device, these fragments were then printed onto glass slides to produce a DNA microarray. The glass slide became a multigene probe and consisted of a grid reference on which each DNA sequence was attached on an 18 mm square area. mRNA expression profile of the yeast was then used to compare cells which underwent a diauxic shift from fermentative growth to aerobic respiration. Within a 21h period, mRNA samples were collected as yeast cells changed their metabolism from glucose utilization to ethanol utilization. These were converted into red-fluorescently labeled cDNA through reverse transcription. The reference cDNA made from mRNA collected at the start of the experiment was labeled with green fluorescence. Hybridization of labeled cDNA to the probe was carried out and subsequent pattern visualized by confocal microscopy. Images of relative intensity were analyzed by a computer to reveal the relative expression levels of the genes. At the beginning of the experiment, all spots are yellow due to the equal expression of mRNA. Later, a red spot indicates a relatively decreased mRNA expression level and a green spot indicates a relatively increased mRNA expression level, compared to the reference.

This study revealed that the diauxic shift resulted in alteration of 28% of genes within a yeast genome. A total of 1030 genes were suppressed and 710 genes were induced by at least two times or more. In addition, altered expression of clusters of genes was also studied, which were shown to have a similar pattern implying that regulation is coordinated by a common promoter element. Such coordinated regulation helps predict roles for ORFs that have unknown functions (Figure 4.2).

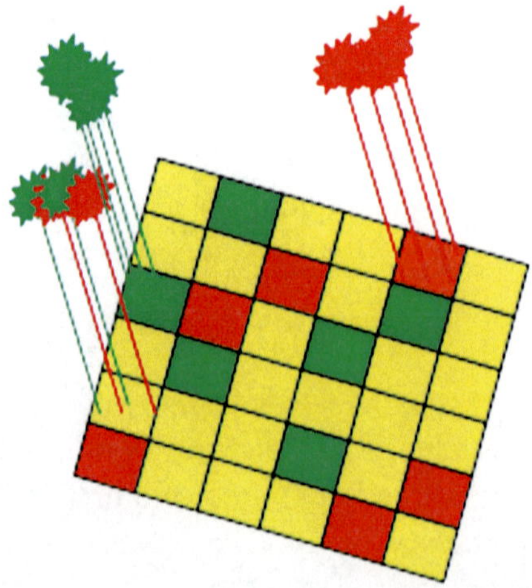

Figure 4.2: DNA microarray technique. 36 of the 6400 ORFs are shown being arrayed on a single slide here. Two samples of cDNA are taken separately; one from yeast cells grown on glucose at the fermentation start are fluorescently labeled with green dye and the other from the cell at later stages of fermentation which are labeled with a red dye. Red fluorescence shows induced expression compared to the reference, green shows repressed expression of genes compared to the reference, and yellow shows no alteration in the level of expression. This array is displaying results from three ORFs. Source: *Curran, B., & Bugeja, V. (2005). Fungal Genetics: A Post-Genomic Perspective Fungi (pp. 65–88): John Wiley & Sons, Ltd.*

For example, few genes were already known to consist of stress response elements (STREs). Through this study 13 additional genes identified to have altered expression levels were examined. It was observed that 9 of these had STRE within their promoter sequence. Therefore, 13 more genes that were previously unknown were now assigned to be linked to stress response.

This technique was also used to identify altered gene expression as a result of the elimination of TUP1 (a co-repressor of transcription) or YAP1 (activator of transcription). This demonstrated that the characterization and dissection of cellular regulation on a genome-wide basis can be feasibly done through DNA microarray technique.

This success was followed by studies on several other fungal genomes. As this method became popular in fungal studies, libraries for cDNA, serial analysis of gene expression (SAGE), alignments, expressed sequence tags (ESTs), macro-, and microarrays, sequence alignments, and BLAST searches all became common rapidly within fungal literature (Table 4.3). *S. Cerevisiae* was the first model organism for analyzing fungal transcriptomics as it was the first available fungal genome and it did not have introns allowing easy amplification through PCR. Other approaches for global analysis of mRNA expression are now available; cDNA, SAGE, and EST analysis.

Next-generation sequencing of DNA has revolutionized all the techniques for sequencing DNA. It includes all the novel technologies which allow recording of sequences of DNA while DNA molecules are being formed in real time. Millions of simultaneous reactions help collect data about DNA sequence which is interpreted by advanced bioinformatics tools. It overcomes the requirement of physical resolving of DNA fragments by electrophoresis which was required in Sanger sequencing. The next-generation sequencing is expected to overcome SAGE and EST analysis. Another advantage of this approach is that does not require DNA probes on the chip as it can directly utilize mRNA as templates.

This technique was useful not only to relate mRNA expression levels in two metabolic states of yeast it was also able to reveal information about their changes with reference to time. This implies that the coordination of gene expression was identified and understanding the effects *when* environmental changes occur and genes are perturbed having an impact on information flow within the cell.

Knowing *when* a gene gets expressed is a significant clue to predict its biological role. For example, highly expressed set of fungal genes during cerebrospinal fluid infection (caused by *C. neoformans)* have been identified using SAGE and microarray technologies. These techniques also helped study the pattern of altered gene expression in *S. Cerevisiae* and Trichoderma reesei resulting in answer to why the glucose metabolism is different in these two species. Moreover, *Aspergillus parasiticus* produce carcinogenic metabolites which were found to be regulated by 18 genes through these techniques.

This means that cellular gene expression pattern, *when* given a particular treatment, helps identify significant factors in biological processes. Moreover, several potential anti-mycotic targets for *Candida* infection have been identified through this approach. DNA microarrays have helped identify *EFG1* and CPH1, two transcription factors, which are induced in *C. albicans* when it is shifted to 37°C while growing in serum. It was established through a transcriptional profile that the pathogenic isolates which were resistant to fluconazole could be grown by inducing resistance in them to evolve in the lab and also to show that itraconazole influences 296 of the fungal ORFs.

Several researches have been successful due to transcriptomics. It was particularly helpful in cases where complicated fungal biological phenomena were under study, both at basic and applied levels. For example, basic research of *S. pombe* revealed that several conserved genes were highly expressed, while a small number of genes that were organism-specific were not expressed highly. This study was possible only due to the presence of transcriptomic data and because searches in databases was available which contained information about ORFs of *S. Cerevisiae*, *S. pombe* and *C. elegans*. Moreover, microarrays also revealed that among 1000 different genes in *N. crassa*, a vast number involved in numerous cellular functions was controlled by its circadian clock. Sometimes, basic research eventually becomes applied research, for example, 62 genes were found to be induced by α-factor in mating type-a cells of *C. albicans* were analyzed using microarrays. Interestingly, seven of these induced genes, previously shown to be involved in virulence, were found to be secretory proteins or cell-surface proteins. Thus, it provided novel targets for developing antimycotic drugs that function by altering the factors controlling the mating pathway. A pure applied research example is of that involving simultaneous expression analysis of a pathogenic fungus, *Magnoporthegrisea*, and its host. The microarray consists of 7137 ESTs of rice and 13666 predicted ORFs of the fungus. Since in appearance in 2005, this microarray has been used for various other studies including identification of a pathogen-specific vacuolar serine protease, Con7 (a transcription factor involved in pathogenicity) – induced genes, genes involved in appressorium formation which helps this species to penetrate into host tissues.

4.3.2. Proteomics

Even though biological analysis in 2-D and 3-D are both very useful, 4-D analysis in time and space is targeted by scientists. This is facilitated significantly by proteomics, which is a lot more challenging than 2-D and 3-D global analysis of nucleotide molecules. Proteomics, in comparison to genomics, involves a group of different but interrelated technologies. 2-D gel electrophoresis enables global analysis of cellular proteins but it is incapable of providing information about their spatial arrangements inside the cell, modifications that occur post-translationally, their structure and function and it is unable to detect proteins expressed at levels below a certain threshold. Several achievements including sequencing of fragments separated on gel and sensitive staining of glyco-specific fragments marked significant improvements in the pre-genomic era. However, the post-genomic era demands more advanced techniques for supplying information regarding all parameters of proteomics to databases.

Table 4.3: Proteomics is much more Challenging than Transcriptomics or Genomics

DNA (information)	RNA (information in time)	Protein (information in time and space)
Nucleotide sequence	Nucleotide sequence	Amino acid sequence
	Level in cell	Level in cell
	Modifications (splicing)	Modifications (glycosylation, phosphate groups, etc.) 3-D structure
		Function
		Location
		Interactions

(Source: *Curran, B., & Bugeja, V. (2005). Fungal Genetics: A Post-Genomic Perspective Fungi (pp. 65–88): John Wiley & Sons, Ltd.)*

Basically, proteomics provides information about the time and levels of expression of genomic ORFs into proteins in the cell. Unlike, RNA, and DNA sequences, proteins cannot be enzymatically amplified and thus limited sample amount is a major constraint in large-scale analysis. However, tryptic protein digestion coupled with chromatography techniques improves detection of presence and abundance of poorly expressed proteins using mass spectrometry (MS). The patterns of peptide fingerprints that are formed due to ionization by peptide fragments through MS, can be compared to similar patterns reported for known proteins in databases. Otherwise, another round of MS can be carried out on individual fragments to reveal information about constituent amino acids. The revealed protein sequences can be searched against databases for both proteins and DNA sequences. Similarly, mixtures of proteins can be run on 2-D gel electrophoresis and individual bands can be collected through excision, followed subsequently by tryptic digestion and mass spectrometry.

This type of proteomic analysis has already been exploited for revealing information about fungal biology. Effects of concanamycin A, Streptomyces-originated antibiotic, on proteome expression in *Aspergillus nidulans* were studied through 2-D gel electrophoresis and MS analysis. Out the 20 observed spots, 5 were excised and sequenced. By searching these revealed sequences against ORF databases, four of them were assigned putative functions. The fifth one was found to have a homolog with unknown function in *A. niger*. In another case, when supplied with flavonoid rutin as the sole carbon source to *A. flavus*, it produces a repertoire of secretory proteins that were analyzed by 2-D electrophoresis. Of the 100 spots formed, 22 were studied further using tryptic digestion, MS analysis and subsequent search against peptide fingerprint databases. Rest of the protein spots (over 90) that remained unidentified were therefore novel proteins for which no data was available on peptide fingerprint databases.

Proteomic analysis, with reference to sequence and abundance, varies from transcriptomic data. On some occasions, proteins are expressed while their mRNAs are not detected. This may be because of a very long half-life of proteins or low production of mRNA. In other cases, mRNAs are found while proteins are present. This may be explained by very quick degradation of proteins, protein expression at undetectable levels, or subjecting of mRNA to regulation post-transcriptionally. Despite these differences, protein analyses are used extensively and on large-scale parallel to transcription analysis for fungal biology. This progress is very quick especially in the case of *S. Cerevisiae*, the species in which cellular protein-protein interactions to form complexes have been studied to some extent through these techniques. With advancements in comparative genomics, information from this model organism is being applied to other organisms.

After detection of proteins and its abundance is found, another important parameter is its 4-D arrangement to reveal its cellular location. To achieve this, several aspects are analyzed including protein's interaction with other proteins, its interaction with other protein complexes and the interaction of these complexes with the cell. The first employed fungal species in this study is *S. Cerevisiae* as it is a model organism which has a highly developed system for gene expression and homologous recombination which allows novel gene constructs to be inserted into specific chromosome loci ensuring appropriate regulation of gene expression. In addition, most of its ORFs are known and characterized.

This was employed by Huh et al in 2003 for studying 4100 ORFs of yeast. He amplified hybrid fragments of these molecules of DNA whose 3' ends of coding regions were fused in-frame with GFP. Subsequently, these hybrid molecules were subjected to homologous recombination in order to insert them back to their appropriate loci on chromosomes. Resulting in the production of 4100 strains of yeast each consisting of a unique DNA sequence having a GFP tag. These proteins can be localized inside cell using fluorescence microscopy. This analysis helped categorize these proteins into 22 locations in cell and therefore 70% of the initially unlocalized proteins (representing 30% of the yeast proteome) were now localized (Figure 4.3).

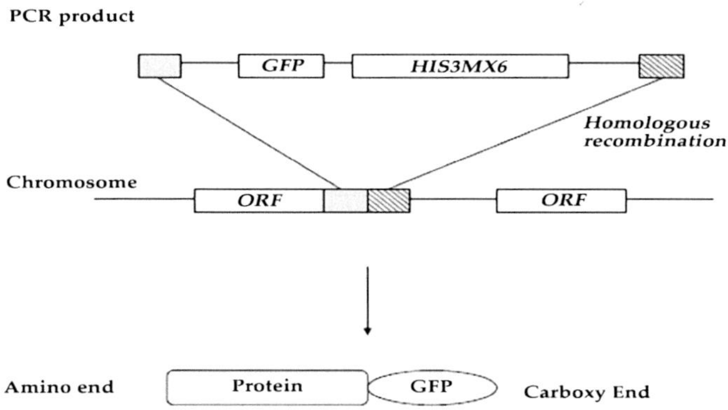

Figure 4.3: Fluorescent labeling of a protein. The target ORF was amplified by creating a hybrid containing His 3 as the selectable marker and gene for GFP fused to the 3' end of the coding sequence. The target hybrid GFP-tagged DNA sequence is then inserted into appropriate chromosomal locus. (Source: *Curran, B., & Bugeja, V. (2005). Fungal Genetics: A Post-Genomic Perspective Fungi (pp. 65–88): John Wiley & Sons, Ltd.*).

This approach was also used for analyzing the yeast's protein interactome. Hue (2003) used a variation of yeast two-hybrid system (Chapter 8) by mating 192 yeast strains producing 'prey' proteins with 6000 yeast strains that expressed 'bait' proteins. The protein interactions in the resultant diploid cells were analyzed using expression of specific reporter genes. Studies using two-hybrid systems provide interactions that are false-positive and is unable to differentiate the weakly interacting complexes. However, they do provide strong indications of protein interactions and functions. This study revealed several insights into proteins' roles in metabolic pathways. These include linkage of two genes with unknown function to metabolism of arginine, a circuit for cell cycle control, a group of protein complexes interacting with a protein significant in ds-DNA breaks, and another involved in meiotic synaptonemal complex formation. Later, a compilation of 2700 protein interactions in *S. Cerevisiae* were discovered, and 1548 proteins in yeast were found to be part of a large network (Figure 4.4). It was noted that proteins with similar cellular functions had strong interactions with each other and hence they belonged to similar functional groups. In this way, around 40 proteins with unknown functions in yeast were assigned putative functions. A huge interconnectedness was found through the global interactome of the yeast proteome. Several proteins present within a specific cluster assigned a central role (such as cell cycle) were found to be interacting with proteins belonging to other classes of clusters (such as transcription, DNA synthesis, etc.). This allowed connection of biological functions to large cell processes. On the contrary, clusters associated to more peripheral functions in cells (such as RNA turnover) were rarely found to be interacting with other functional clusters. Overall, a total of 1548 yeast proteins along with their 2700 links were categorized into a network consisting of 32 functions with around 70 connections. This complex protein interactome is represented graphically in the form of protein-interaction maps.

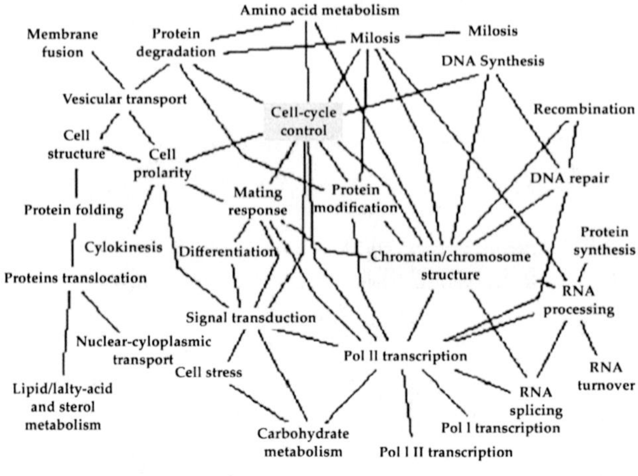

Figure 4.4: The yeast proteome meta-network. It consists of 1548 types of proteins having 32 functions and 2700 links between them. (Source: *Curran, B., & Bugeja, V. (2005). Fungal Genetics: A Post-Genomic Perspective Fungi (pp. 65–88): John Wiley & Sons, Ltd.*).

This method helps study interactions of proteins that occur when the proteins are induced to express highly and constitutively and allowed to interact in the same compartment of cells at the same time. A much more powerful analysis of this interaction was carried out by Gavin in 2002. He generated hundreds of affinity-tagged ORFs of yeast with their normal regulatory regions and transformed them into cells to allow homologous recombination. Expression of these cellular proteins was induced followed by isolation of their complexes using tandem affinity purification (TAP). It allowed elusion of proteins with their interacting proteins that were identified using MS analysis. Through this method, 232 protein complexes and 1440 separate proteins were identified in yeast. It helped discover 231 proteins that were previously uncharacterized and 113 which had no known cellular function. Several yeast global analysis help to complement and support these studies including:

1. Analogous to DNA microarrays, protein microarrays is a technology in which target proteins are immobilized on to a solid support which is probed with proteins that are fluorescently labeled. The proteins that are bound are therefore expected to interact *in vivo*;
2. A process for genetic screening, synthetic lethal screening analyzes pairs of mutations together. It helps identify mutations that are lethal when together but not when alone. This implies that both the proteins have some sort of in vivo functional interaction;
3. mRNAs that have similar regulation indicates that the proteins formed from their translation may be involved in similar cellular processes.

High-throughput systems that help in studying protein-protein interactions are shown (Figure 4.5). Conclusively, these techniques provide much deeper and more dynamic insights into biological processes than before. Moreover, comparison of various model organisms implies that these processes are conserved evolutionarily since a long time as they are found in millions of organisms. Information about interactions of proteins in one organism can be extrapolated to predict its role in other organisms. It is, however, necessary for scientists to communicate properly as different methods are used for naming and describing genes and their products in different organisms. This is the reason that as long ago as 1998, scientists facilitated communications across various databases by using controlled common shared vocabulary, and a defined framework for organization of information revealed from study of the only three eukaryotic genomes that were available at that time. This was called the Gene Ontology (G) project. Today, this project is the key to the post-genomic era as it allows standard annotations, computational queries, data recording, retrieving, and analyzing from multiple sources.

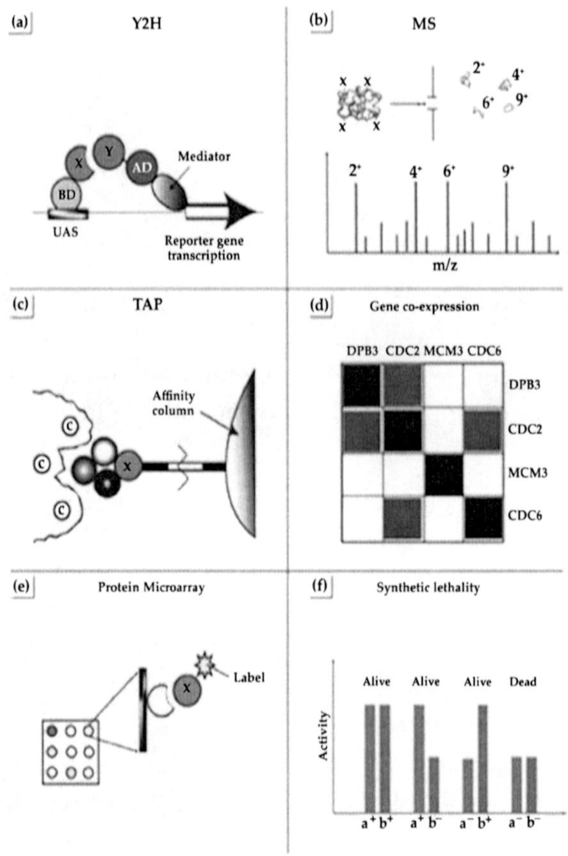

Figure 4.5: Protein-protein interactions studied through high-throughput systems. A) Yeast-two-hybrid system; shows linkage between proteins X and Y. X consists of a domain that binds to DNA sequence of a promoter (UAS). B) MS; analysis of polypeptide sequences C) TAP; proteins associated with target protein are purified. (Source: *Curran, B., & Bugeja, V. (2005). Fungal Genetics: A Post-Genomic Perspective Fungi (pp. 65–88): John Wiley & Sons, Ltd.*).

Ontology is based on graphical presentation where a map of nodes (consisting of words from a controlled vocabulary) connected by arrows (showing relationships between terms based on logical statements) are developed. As mentioned in section 3.3, the vocabulary in gene ontology (GO) is used to describe three aspects of molecules; their function which includes vocabulary for defining their gene products' biochemical activity (e.g., enzyme, ligand etc.), their cellular locations which explains the location inside cell where they are active (e.g., nucleus, cytoplasm, etc.), the terms describing their biological processes in which they participate (e.g., mitosis, signal transduction etc.). An example of a low-resolution map for biological processes is shown (Figure 4.6). The nodes have the terms used to describe the biological processes and the connecting arrows represent their logical relationship as " is a" or "is part."

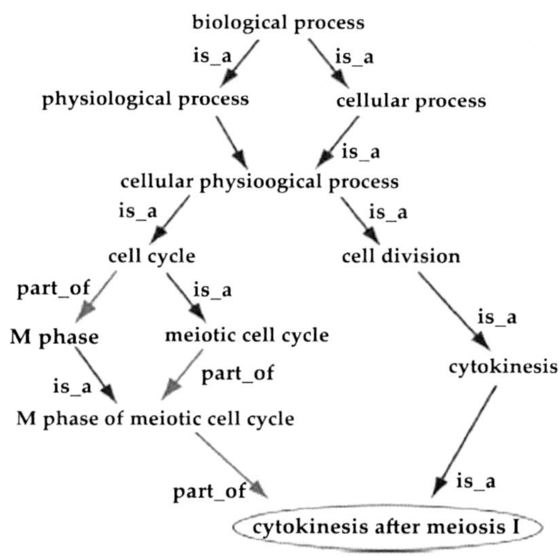

Figure 4.6: A low-resolution map showing biological process ontology. It is made using terms from a strictly controlled vocabulary describing biological processes which are linked to logical statements. It starts from top and reaches the bottom. (Source: *Curran, B., & Bugeja, V. (2005). Fungal Genetics: A Post-Genomic Perspective Fungi (pp. 65–88): John Wiley & Sons, Ltd.*).

It should be noted the gene ontology project in not process for naming genes and their products, rather it is used to query and retrieve genes and proteins. GO describes the attributes of the biological molecules (their locations, function, and processes) and not the molecules themselves. However, the association of a GO term to the genes or gene products stored in various databases like EMBL, Genbank enables access to the molecules themselves. This association should be evident from literature review and the date and the type of evidence also has to be presented in the reference (e.g., microarray expression, mutant phenotype, etc.).

One GO term in, for example, *molecular function* ontology is therefore linked to various genes in multiple organisms in different databases. Interestingly, novel gene sequences, based on sequence similarities, can be matched with GO terms associated to annotated sequence in databases. Through this process, annotations from the well-characterized and annotated *S. Cerevisiae* have been transferred to fungal species that were poorly characterized, such as the fungal pathogens *Pneumocystis carini* and *Sclerotinia sclerotiorum* etc. Fungal biologists can, therefore, apply this project to understand fungal biology without the need for highly developed technologies.

Even though *S. Cerevisiae* is highly characterize, to study 4-D arrangement of all cellular proteins and its interactions with their biochemical kinetics is still not possible. However, theoretical models representing interactions within groups of elements involved in specific biological tasks have been developed.these models can be extended for developing a framework for molecular experiments that produce data which in turn helps refine the model. *Systems biology* is an approach that reintegrates molecular studies to uncover insights about cellular processes in which they participate.

4.4. STUDIES INVOLVING SYSTEMS BIOLOGY

This field helps provide insights into biological phenomena. For example, glucose metabolism in *S.cerevisiae* has already been studied extensively using traditional methods. It has helped find nine genes that are induced in the presence of galactose and in the absence of glucose. Their gene products include one transporter proteins, four transcription factors and four enzymes whose metabolic substrates and regulatory mechanisms have been elucidated.

All this available information was first used for defining a model for glucose utilization in step 1 is system biology. In step 2, 20 distinct cell conditions were yielded by systematic perturbations of each component of the pathway. Cells that were wild-type were therefore developed followed by production of deletion strains, each missing one of the nine genes, in the presence or absence of galactose. Microarray technology was utilized to globally detect and quantify cellular responses in each strain. Additionally, protein expression was also analyzed in wild-type cells. Step 3 of this approach is carried out by integration of the initial model with the newly available information and with the reported information about yeast protein-protein and protein-DNA interactions. Interestingly, DNA microarray showed that expression 997 ORFs from yeast were altered in one or more of these perturbations even though only nine gene products are required for metabolism of galactose. It revealed that the 'great interconnectedness' of yeast proteome is highly perturbed by a few perturbation within galactose metabolism. The researchers hypothesized explanations for unexpected observations in step 4. This was further extended by inducing further perturbations and subsequently testing the newly devised model. Obtaining these results was possible only due to sequenced genomes of yeast, developed mRNA analysis for this species, data for its protein-protein interaction, detection of protein presence through MS analysis and due to easy access to computational and all other facilities.

4.5. CONCLUSION

Rapid progress has taken place in fungal biology from one-one enzyme in *Neurospora* to yeast cell cycle. The potential for fungal system in this post-genomic ear is much more useful compared to the previous eras. A rapid shift from *S. Cerevisiae* as the model fungal organism to various other organisms has taken place due to enormous comparative biology. A rapid progress in a diverse range of fields is possible due to this approach, including, exploration of native and transgenic fungal biology, identification of genomic differences to achieve improvement in strains, understanding of molecular alterations due to fungal infection and identification of new anti-mycotic targets. Therefore, the post-genomic era offers a limitless range of research opportunities.

5
CHAPTER

FERMENTATION IN FUNGI AND THEIR PRODUCTS

CONTENTS

5.1. INTRODUCTION

Several fungi have long been used by humans for production of nutritious (bread), enjoyable (alcohol), and medically-significant (alcohol) products. Environmentally, fungi are important for decomposition of organic compounds. These decomposers produce commercially-important antibiotics and enzymes. Since thousands of years, yeast has been utilized for production of alcohol and bread as early as 6000 BC. Microscopic analysis of yeast was done in 1680 and specifically, *S. Cerevisiae* was visualized in 1838. Fermentation process in yeast has been exploited even though its mechanism was elucidated in 1863 by Loius Pasteur. Historically, fungi have been both advantageous and disadvantageous to humans. One of the major discoveries was that of penicillin by Alexander Fleming as the healer for wound infections.

Table 5.1: Fungal Applications at Industrial Level

Process	Example
Biomass	Production of baker's yeast Brewer's yeast tab- lets Single-cell protein (mycoprotein) Proteins (native or recombinant)
Cell components	Antibiotics (penicillin) Vitamins (B_{12})
Catabolite products	Ethanol
Bioconversion	Carbohydrates break down

(Source: *Kavanagh, K. (2005). Fungal Fermentation Systems and Products Fungi (pp. 89–112): John Wiley & Sons, Ltd.*)

5.2. FERMENTATION SYSTEMS OF FUNGI

Fungi have been used for industrial production of a wide range of foods, such as bread and mycoprotein, alcoholic beverages, such as beer and wine, recombinant proteins, antibiotics, and vitamins (Table 5.1 and 5.2). Depending on the fungal nature (yeast or filamentous), product type, and the production scale, various fungal fermentation systems have been employed. Yeast cells have a population doubling time of around 1.15 to 2 h, whereas, yeasts have 2 to 7 h time, therefore rendering the latter to have limited application as a fermentation system. Additionally, yeast is more morphologically favorable for use as fermentation systems. Some basic fungal systems have been developed as fermenters which are modified to achieve fine-tuning in order to achieve production of various products of fungi.

Table 5.2: Products Obtained from Various Fermentation Systems of Fungi

Fermentation Type	Examples	Product
Solid	Cultivation of *A.bisporus*	Mushrooms
Cell-recycle batch	Winemaking	Summer wines
Fed-batch	Cultivation of *S.cerevisiae*	Baker's yeast
Batch	Brewing, winemaking	Beer, wine
Continuous	Cultivation of *Penicillium graminerium*	Mycoprotein

(Source: *Kavanagh, K. (2005). Fungal Fermentation Systems and Products Fungi (pp. 89–112): John Wiley & Sons, Ltd.*)

5.2.1. Solid Fermentation

This type of fermentation occurs when fungi are grown without free water on a solid substrate like grain. The species grown should be able to survive in low water levels. Such systems have been used for growing *Agaricus biporus*, the edible mushrooms and filamentous fungi for production of antibiotics and enzymes. Solid fermentation of *Rhizopus* has been used for producing soy sauce from soybeans and that of Penicillium sp. for production of cheese from milk curd. Moreover, solid fermentation systems have been used for horticulture compost production. Fungal biomass can be produced by degrading straw by simultaneously employing yeast (*Candida lipolytica*) and filamentous fungus (*Chaetomium celluloyticum*). Even though solid fermentations are hard to control and expensive to sterilize, it has several applications.

5.2.2. Batch, Continuous Cycle, and Immobilized Fermentation Systems

This type of fermentation involves an initial low inoculum of fungal cells. When allowed to grow, the cells reach the lag phase in which biochemical cellular activities are intense which continues for minutes or hours. The exponential phase follows in which nutrients are maximally utilized unless they start to decline. This decline in nutrient levels slows down the rate of growth of the population of fungi until they reach the stationary phase. During this phase, the population of cells becomes constant. If incubated for a longer time, cell viability decreases and the cell number reduces. When strains of brewing yeast are used, they produce maximum ethanol during their stationary phase. When the levels peak, brewers carefully monitor the appearance of ethanol and the fermentation is stopped. Prolonged fermentation results in usage of ethanol as replacements for carbohydrates. Therefore, a biphasic growth curve is shown by the culture. Similar to batch fermentation, the fed-batch can also be used. In this case, nutrients are periodically supplemented during the process of fermentation. This system maximizes biomass production and is used for producing baker's yeast/ Cell recycle batch fermentation (CRFB), also called as Melle-Bio not process is another type of batch fermentation. In this system, a high density of cells in ensured throughout the process by inoculating the next fermentation medium by cells present at the end of previous fermentation reaction. Maintenance of this high cell density increases the available energy for production of ethanol and improves adaptation of yeast cells to tolerate elevated levels of ethanol. This type of fermentation continues for weeks or even months, each time initiating a new cycle by inoculation with cells from the previous cycle. This process is specific to strain of yeast and substrate type used. It has been employed for producing summer wines in Italy and for producing ethanol using the yeast, *Pachysolen tannophilus* from xylose as the carbon source (Figure 5.1).

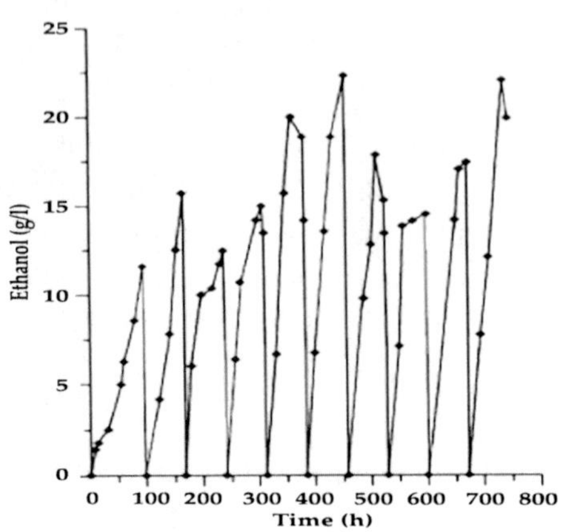

Figure 5.1: Cell recycle fermentation. The yeast *Paschysolen tannophilus* is used for ethanol production using xylose. (After each fermentation cycle, yeast

cells are collected and inoculated into fresh fermentation medium. (Source: *Kavanagh, K. (2005). Fungal Fermentation Systems and Products Fungi (pp. 89–112): John Wiley & Sons, Ltd.*).

Enhanced tolerance to ethanol is accompanied with increased ethanol production in the latter case. When the same species is used for fermentation from glucose, there is no significant effect on ethanol through CRBF.

In continuous cycle fermentation system, the cells are allowed to grown in steady conditions and maintained in a particular growth phase. This is achieved by continuously supplying the medium with nutrients and biomass, and also by periodic removal of used medium in order to maintain the total volume of reaction. Even though such cultures are difficult and expensive to maintain they are often utilized for production of mycoproteins and antibiotic.

To optimize productivity, fungal fermentation systems can also be immobilized. This methods lowers the chances for contamination, removes the toxic metabolites which could inhibit cell growth, and helps maintain high density of cells. However, a disadvantage of this method is the reduction in viability of fungal cells and the degradation of immobilization systems due to over-usage.

Immobilization of fungal systems can be done through four different ways. The fungal cell can be placed inside membranes, like microfilters, dialysis tubing or other porous material (Figure 5.2 a). In this type of system, membrane can become clogged reducing the effectiveness of this system.

Alternatively, the yeasts cells can be electrostatically or covalently attached to solid surfaces (Figure 5.2 b). Cells can be cross-linked using, for example, glutaraldehyde, and then attached to a solid support which can either act as a substrate over which growth medium can flow or it may be immersed completely into the growth medium.

Alginate beads can also be used to incorporate fungal cells (Figure 5.2 c). a 2–4% sodium alginate solution is prepared and the cells are suspended in it. This is followed by dropping of beads into calcium chloride solution which results in the formation of spherical beads consisting of insoluble calcium alginate. Porous, inert, and stable beads that contain fungal cells are formed as a result. This method is very useful for ethanol production despite the disadvantage of a collection of gases (like CO2) which may damage the beads. Alginate can also be replaced by agar, polyacrylamide, and gelatin for entrapping yeast.

Yeast cells can also be immobilized by the process of flocculation or clumping. At the end of fermentation, flocculation occurs naturally. When induced towards the conclusion of fermentation a high density of cells is ensured which eliminates the need for centrifuging cells to harvest cells.

Entrapment of yeast cells within a membrane (Figure 5.2 a) offers several advantages such as; maintenance of high density of cells, increased the productivity of ethanol, and carry out dilution at high rates within a continuous cycle system. In the entrapped fermentation system, yeast metabolism may fluctuate influencing the nature and quality of the product. Additionally, the metabolites are prevented from diffusing into the surrounding matrix due to the presence of membranes. This can have negative effects on cell viability within the matrix if toxic products are formed which are unable to diffuse out easily.

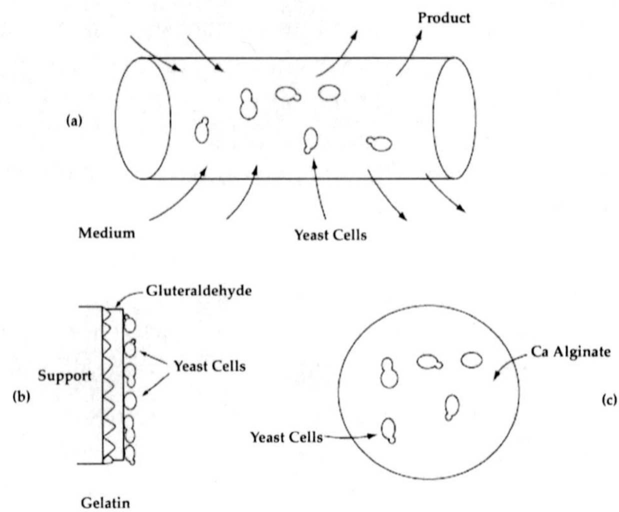

Figure 5.2: Methods for immobilization of fungal cells. (Source: *Kavanagh, K. (2005). Fungal Fermentation Systems and Products Fungi (pp. 89–112): John Wiley & Sons, Ltd.*).

Regardless of the fermentation system employed, the products formed require to be concentrated before they are used. Usually, lipids and antibiotics are produced at concentrations of 10–30g/L, single-cell proteins at 30–50 g/L and ethanol a 70–12 g/L. to ensure reliable and continuous recovery of products, downstream processing (DSP) is needed. This can be done in various ways depending on the product nature. For example, distillation is used for whiskey production while protein isolation requires sonication, degradation of the cell wall by enzymes or use of French press for compression. The nuclease is used to degrade nucleic acids released by lysis of cells. Ammonium sulfate precipitation helps in collection of enzymes which can be further purified using columns made up of DEAE-Sephacel, Sephadex or hydroxyapatite. Once purified, other agents may be added into the product to make it ready for marketing.

5.2.3. Factors influencing Productivity of Fungal Fermentation

Parameters affecting the fermentation systems vary depending upon the type of the chosen system which in turn depends on the requirements of specific processes. For example, ethanol production is high using batch fermentation, whereas, it would be poor when film fermentation is used. Type of fungi used is also an important consideration. For example, a bioreactor uses spargers (paddles) to allow efficient stirring of nutrients and transfer of oxygen. Yeast cells are more likely to grow in bioreactors compared to filamentous fungi which propagate by the production of mycelium (which is disrupted by spargers). Instead, fungal mycelium is agitated in a bioreactor through air or gas bubbling within the medium (Figure 5.3). Fermentation systems differ in their requirements for oxygen, depending on whether the fungi carry out aerobic or anaerobic respiration. Sometimes in the case of filamentous fungi, the optimal growth rate may require a specific inoculum initially. Optimization of the fermentation medium is another factor influencing fermentation. For example, some fungi can become inhibited by catabolites, therefore, increased concentrations of glucose may inhibit the process. Temperature is another major factor that affects fermentation systems' productivity. Temperature can, for instance, determine the production of either ale or lager while brewing.

Of all the available fermentation systems, batch or fed-batch are most prevalently used ones. This is because they are easier to set up and in various sizes according to the demand. Batch fermentation is employed for producing beverages. When the yeast cells are used for brewing, at the end of each fermentation cycle, they are discarded and a fresh inoculum is used. This ensures that during prolonged fermentation, characteristics specific to a strain which are associated to a particular beer are not lost. This is very significant in maintaining consistency of beers.

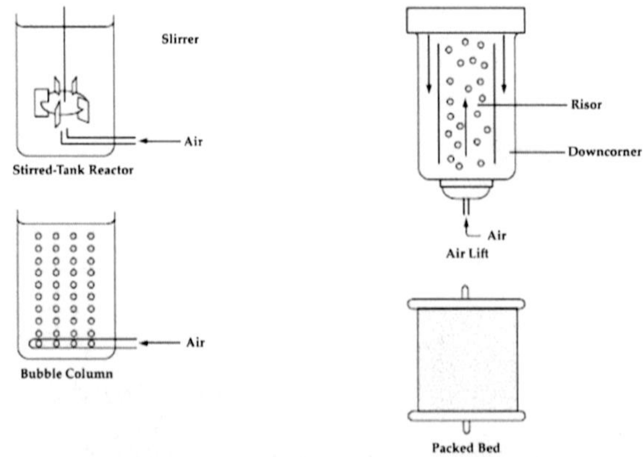

Figure 5.3: Methods for aeration of fungal fermentation mediums. (Source: *Kavanagh, K. (2005). Fungal Fermentation Systems and Products Fungi (pp. 89–112): John Wiley & Sons, Ltd.*).

5.3. COMMERCIAL PRODUCTS FROM FUNGI

5.3.1. Brewing

Sugars can be fermented by yeasts in semi-anaerobic conditions in order to produce alcohol for the manufacture of beer, wine or cider. Along with ethanol, organoleptic compounds are also formed by yeast fermentation which adds flavor to the product. These include ethylene acetate, butanol, and isoamyl alcohol. During fermentation, glucose is converted into pyruvate by yeast through the glycolysis process. This pyruvate is changed into acetaldehyde with the aid of enzyme pyruvate decarboxylase releasing carbon dioxide. Acetaldehyde is finally converted to form ethanol by the help of alcohol dehydrogenase. Overall reaction is:

$$C_6H_{12}O_6 \rightarrow 2C_2H_5OH + 2CO_2$$

Simple sugars like glucose or fructose in grapes and apples are fermented to produce wine and cider, respectively. For beer production, the carbon source starch comes from barley which is not fermentable by yeasts as they lack amylase enzymes. This poses a little complication in the process. Barley is therefore germinated by the process of malting. During this process, starch is converted into fermentable sugar groups which are utilizable by yeast for ethanol production. This is done by employing endogenous enzymes, α-, β- and glucoamylases (Figure 5.4). The β-1,4 glycosidic bonds within starch are broken down by α-amylase to yield oligosaccharides containing 7 to 12 glucose molecules. Maltose units are released from non-reducing ends of these chains by β-Amylase. This process of degrading starch into dextrins and maltose is called as saccharification. Following malting, the seeds are dried through different ways. The nature of drying affects the type of beer that is formed at the end. If drying is long and cool, malt produced has high activity of enzymes and is 'pale.' On the other hands, upon hot and short drying procedure, the malt produced has low activity of enzymes and is 'dark.' Drying is followed by the crushing of seeds through milling which produces the 'grist.' To obtain sugars and enzymes, hot water is sprayed on to it. At this stage, most of the sugars in the 'wort' are non-fermentable. Upon amylase action for 2 to 3 hours, they get converted into fermentable sugars which constitute 75% of the wort. Typically, at this stage, fermentable sugars in wort at this stage include: glucose (9.1 g/L), fructose (2.1 g/L), maltose (52.4 g/L), sucrose (2.3 g/L), and maltotriose (12.8 g/L) and the nonfermentable carbohydrates are present at 23.9 g/L concentration. Once formed, the wort is sterilized and its pH reduced by boiling. To improve flavor of the beer, hops are included in the wort during boiling.

The wort is then allowed to cool and is filter. Yeast is inoculated to induce fermentation process. The commonly used yeasts for fermentation typically had optimum temperatures within 15 to 22°C range. At the end of the process, they got risen to the top of the beer and therefore, were called as top-fermenting yeasts. On the other hand, the bottom-fermenting yeasts are now used for production of most of the beers and these have comparatively lower optimum temperatures. At the end of the fermentation process, they sink to the bottom of the beer through the process called flocculation.

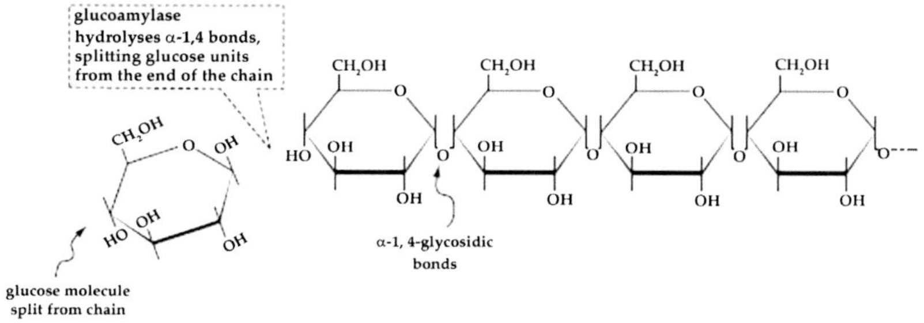

Figure 5.4: Starch is degraded by amylase into numerous fermentable and non-fermentable sugars. (Source: *Kavanagh, K. (2005). Fungal Fermentation Systems and Products Fungi (pp. 89–112): John Wiley & Sons, Ltd.*).

Brewing can be regulated through agitation, and inoculum size and temperature changes. Depending on the nature of the barley used (pale or dark), the flavor of beer can vary. In addition, these are affected by hops and ionic compositions, nature, and amounts of organoleptic compounds. Nature of the beer is altered due to ionic composition. Compositions of, for example, sodium and carbonate vary the taste of the final product. The end product of brewing is filtered to allow removal of yeast, followed by pasteurization and packaging before dispatching. However, some beers are left to allow maturation in the final step.

A wide variety of beers are available. Ale is produced when fermentation takes place at 20–25°C for 36-h following which, cooling period of 72 h is carried out at 17°C. Lager fermentations, on the other hands, require lower temperatures in the range of 7–11°C for a period of 12 to 14 days. Some beers need an extra amylases source like rice. "native beers' have been produced using human salivary amylase from millet or sorghum. The grains are first grinded and then incubated with amylase which breaks down starch into sugars fermentable by yeast. When exogenous amylases and glucoamylases are employed for converting non-fermentable carbohydrates into fermentable sugars, 'light beers' are produced.

Even though traditional methods are still popularly used for production of alcohol, attention has been given for improvement of yeast in order to enhance ethanol yield, reduce time for fermentation and to enhance flavors of beer. Generation of brewing yeast that is capable of converting α-acetolactate into acetoin has been successful through genetic engineering. This eliminates the need for an organoleptic compound, diacetyl, which cause off-flavors. This conversion is induced by addition of a bacterial gene for the enzyme α-acetolactate decarboxylase (α-ALDC) into the yeast. Even though low levels of diacetyl have been detected when this yeast is used for fermentation, it is not used commercially yet. Moreover, genetic engineering of brewing yeasts has helped in achieving production of amylase and glucasae, and frequent flocculation. Productivity and efficiency of ethanol have been achieved through variations in brewing process. To produce ethanol, immobilized yeast cells have been used to inoculate continuous cultures, which improves the cell density attained, reduces the diacetyl production and speeds up the fermentation process. Due to controversy in using genetically modified produces, the commercial application of genetically improved has not yet been done even though it offers various advantages over traditional yeast strains.

5.3.2. Bread-Making

The yeast *S. Cerevisiae* plays a key role in making bread. It is mixed with, aerated and allowed to stand for a short time period in a warm place. During this period, yeast respires to produce carbon dioxide (Equation shown below).

C6H12O6 + 6O2 → 6CO2 + 6H2O

Trapping of carbon dioxide gives the break a light and airy texture. During baking, aerobic process occurs. An insignificant amount of semi-anaerobic conditions developed within the dough may also result in formation of ethanol. It gets burnt off along with the carbon dioxide during baking at high temperatures resulting in 'fruity' scents linked to bakeries. Some countries like Japan produce bread that essentially contains ethanol. This is done by leaving the dough for longer periods of time to allow using up of oxygen in oxygen and commencing of fermentation.

5.3.3. Single-Cell Protein: Food from Fungi

Single-cell protein (SCP) can be yielded using a wide variety of fungi (Table 5.3). *Trichoderma viride* is a filamentous fungus which helps obtain SCP using cellulolytic material. Other the other hand, *Saccharomycopsis fibuliger,* and *Candida tropicalis* yeasts use starch during the 'Symba process' to produce SCP. Starchy effluents produced from food-processing factories are used for inoculation of *S. fibuligera.* This yeast hydrolyzes and ferments the solubilized starch. The resultant oligosaccharides and reducing sugars are then employed for growth of C. tropicalis which is used as a dietary supplement.

Producing SCP from fungi offers several advantages including:

1. Fungi are organisms that are well-characterized;
2. It is acceptable to consume fungal organisms (e.g., mushrooms) and their produce (e.g., cheese);
3. The nucleic acid content in fungi is low;
4. They have a filamentous morphology which can be used to provide a fibrous texture and appearance similar to meat;
5. Fungi have been used to produce SCP as a commercially available food.

Table 5.3: Materials and Fungal Species Utilized for Production of SCP

Material	**Fungal Species**
Cellulose	*T.viride*
Ethanol	*Candida utilis*
Banana peels	*Pichia* species
Beef fat	*Saccharomycopsis* species and *C.utilis*
Sugars	Range of yeast species
Starch	*Saccharomycopsis fibuliger*

(Source: Kavanagh, K. (2005). Fungal Fermentation Systems and Products *Fungi* (pp. 89–112): John Wiley & Sons, Ltd.)

Mycoprotein is synthesized by *Fusarium graminearum under the trademark* Quorn™ mycoprotein. Rank Hovis McDougal in 1968 began the search for fungal isolates that produce mycoprotein in large quantities. He studies 3000 strains obtained from soil which were able to produce proteins when grown on starch as the carbon source. Initial studies were done on Penicillium sp. But later F. graminearum was found to be superior. This isolate (A3/5) could grow optimally at 30°C with a 3.5h generation time. When 54g of fungus is provided with 100g glucose, 45% of the obtained weigh is for proteins.

Fermentation is currently carried out in large containers that have up to 3000 g cycles times. Cultures containing starch are maintained at exponential growth phase of fungi. A disadvantage of such a long fermentation time is that a variant appears which is highly branches and it has a huge negative impact on product's texture. Formation of this variant can, however, be inhibited by providing lesser nutrients and by regulation of pH.

Harvest of fungal mycelium is carried out regularly; it is then treated with heat in order to lower the content of RNA from 10% to 2% and then further processed for marketing. This reduction is recommended because consumption of RNA has harmful effects on the body. It can cause accumulation of uric acid which gets deposited in joints and kidneys. The net protein utilization (MPU) of fungal mycoprotein is 75, compared to the NPU values for beef and fish, which are 80 and 83, respectively. It can be processed to make its texture and appearance similar to meat.

The Quorn mycoproteins are sold in the form of cubes as ready-to-eat meals, which are flavored to be used similar to chicken. Because this protein is not originated from animals, it has low levels of fat and cholesterol. Therefore, it is healthier and suitable for vegetarians.

5.3.4. Production of Antibiotics

The first fungal antibiotic was discovered by Alexander Fleming in 1928. He found that a substance was released by *Penicillium notatum* which inhibited growth of gram-positive bacteria. This chemical was named penicillin, classed as a β-lactam containing antibiotic. They work by preventing synthesis of peptidoglycan in gram-positive bacteria.Later, a screening was done for different isolates and *P. chryosgenum* was found to be the best producer of this drug. After some selection and mutagenic procedures, the conventionally used strain was able to produce penicillin at 7000 mg/mL rate, compared to the first identified isolate that produced at 3mg/L. Currently; submerged fermenters are used rather than solid fermenter. Three variants of penicillin are formed through fermentation; V, G, and F. Type V are ready to be used as it is formed while G needs some modifications to help it resist the action of bacterial penicillinases. These enzymes are produced by bacteria to kill the fungal cells.

Most fungal antibiotics are secondary metabolites formed as products of fermentation. This production is associated with sporulation and happens at the stationary growth phase. Antibiotic production is inhibited through catabolite repression, which can be prevented by lowering glucose amounts in the medium or by mutating fungi in order to induce catabolite repression resistance. Elevates nitrogen and phosphate levels are also inhibitory to the production of antibiotics. Sometimes, elevated levels of antibiotics in the fermentation medium can retard its own production of damage the fungal cells, hence resulting in feedback inhibition. This effect can be reversed using ampothericin B, an antifungal agent. It works by making their cell membrane permeable so that level of antibiotic present intracellularly is reduced, thus increasing its production.

Productivity of antibiotics can be enhanced through selections, mutations, or by allowing mating between a strain that is a good secretor and a strain that is capable of producing the antibiotic. Additionally, rational selection allows complexing of all metal ions in medium by introduction of a chelating agent, thereby, increasing antibiotic production. Genetic engineering can also be employed for introducing genes responsible for antibiotic production into other species. It also provides a possibility to produce hybrid antibiotics having novel targets.

5.3.5. Production of Enzymes

Many fungi, mostly belonging to the genera *Aspergillus and Penicillium notatum* are used for production of industrial enzymes. These are summarized in Table 5.4. They are popularly used since they are well-studied organisms that are safe for use.

Table 5.4: Sources and Applications of Enzymes Produced by Various Species of Fungi

Fungal Species	Enzyme	Uses
Yeast		
S.cerevisiae	Alcohol dehydrogenase	Ethanol assay
S.cerevisiae	Invertase	Confectionery
Kluyveromyces lactis	Lactase	Dairy industry
Mould		
A.niger	β-Glucanase	Brewing
A.oryzae	Protease	Meat tenderizer
A. oryzae	α-Amylase	Food industry
A.niger + A.oryzae	Lipase	Dairy industry
Aspergillus species	Pectinase	Fruit juice clarification
T.viride	Cellulase	Dehydrated foods
Mucor miehei	Rennin	Cheese manufacture

(Source: *Kavanagh, K. (2005). Fungal Fermentation Systems and Products Fungi (pp. 89–112): John Wiley & Sons, Ltd.*)

Usually, solid fermentation systems are used for culturing of fungi for enzyme production. The commonly used substrates are grains such as bran. More recently, bed batch, continuous cycle or submerged systems have also been employed for this purpose. There are several advantages in using fungi as enzyme sources, including:

1. Fungal metabolism is more flexible and variable compared to other animal's metabolism;
2. They are easy to grow and genetically manipulated;

3. Specific enzymes can be produced through mutagenesis and by employing selection pressure;
4. To produce recombinant proteins, fungal transformations can be successfully done using several vectors;
5. Maintenance of fungal enzymes in a reactor is possible by immobilizing them. It increases their effectiveness by improving their stability and activity by altering pH.

The major problem encountered in using fungal fermenters for large-scale enzymes production is the difficulty to regulate the temperature of the bioreactor. Moreover, industrial enzyme production is affected by fermentation for longer periods. This is because of loss of advantageous characteristics specific to fungal strain. Once cultured, the fungal cell wall needs to be degraded to release the intracellular enzymes. This is usually done by using digestive enzymes, treatment with alkali, or by mechanical damage. When released, the enzymes have to be purified to remove cell debris. This is done using affinity chromatography, both in immobilized or free state.

5.3.6. Mushrooms

Mushrooms are popularly consumed worldwide. They are produced in largest quantities in China, USA, and Netherlands, with a yearly production of 442,000 t, 358,000 t and 240,000 t respectively. UK are the largest importers of mushrooms whereas export to Europe is chiefly from Poland and Netherlands. *A. bisporus* is the edible species of mushrooms most prevalently eaten in Western Europe. On the other hand, Far East Europe cultivates species including *Auricularia*, *Pleurotus*, and *Lentinusedodes* (Figure 5.5).

Compost is used as the cultivation medium for *A. bisporus,* which is composed of straw, stable manure and chicken manure. The fungus cultivated on grains of cereal is mixed with compost. The mycelia of the *A. bisporus* mushroom are allowed to grown on compost at around 25°C under humid conditions for up to two weeks. These mycelia are made up of septate hyphae rich in chitin. They produce extracellular digestive enzymes releasing simple nutrients that are eventually absorbed into the fungal hyphae to allow its growth. Following this stage, the fungi are stimulated to develop primordia which eventually become mushrooms. This is done by 'casing' neutralized peat over the compost. This casing likely provides factors (chemical, physical or/and biotic) that influence formation of mushrooms. Ideally, weekly appearance of flushes or crops is observed unless the nutrients in the compost are used up.

Growing mushrooms in these conditions can also allow development of other harmful fungal species. These can be either weed molds or fungal pathogens. The former are competent with mycelium of mushroom for water and nutrients in the compost and for growth space. The latter, on the other hand, affects the mushrooms directly.

Species like *Coprinus*, *Chrysporium*, and *Trichoderma* influence fungal growth and yield variable. The green mold (*Trichoderma*) grows on the casin, whereas, *Chrysporium retards mushroom mycelium growth by growing in the interface between casing and compost.*

Lecanicillium fungicola, *Mycogone perniciosa,* and *Hypomyces rosellas* are some examples of pathogens of mushrooms. They cause diseases that are very harmful and they reproduce by spores formation which are spread through contact with air, water, and humans. *Lecanicillium fungicola* causes dry bubble in which stipe of the mushrooms swells up and grey spots appear on the cap from where development of mycelium may initiate. *Mycogone pernicosa causes* wet bubble which causes infection on artificially-grown mushrooms along with wild mushrooms and toadstools. When humidity levels are high, stripes of the mushroom are swollen and their caps are irregular and small. The mushroom surface is covered with droplets that are amber in color which eventually turns into a decaying mass like a wet bubble. *Hypomyces rosellas* causes the cobweb disease which inhabits the inside of mushrooms and also over and through the casing. Mycelia that are cob-web like engulf the mushrooms and their gills are unable to develop.

Figure 5.5: a) Mushroom showing evidence of *Lecanicillium* dry bubble infection. b) *Trichoderma* inhibiting the growth of mushrooms. (Source: *Kavanagh,*

K. (2005). Fungal Fermentation Systems and Products Fungi (pp. 89–112): John Wiley & Sons, Ltd.).

These fungal pathogenic species account for around 90% of the diseases in mushrooms rendering a huge loss in economy to mushroom cultivators. Careful hygiene and fungicidal use can prevent their growth.

5.3.7. Soy Sauce

Soy sauce, a common ingredient in Asian food, is also produced with the help fungi. Crushed wheat and soybeans are used for solid fermentation to produce soy sauce. The soybeans are first cooked to allow removal of contaminants and subsequently allowed to mix with wheat grains. This mixture is pressed into cakes on which the inoculation of *A. oryzae* is done. Fermentation is allowed to continue for 1 to 2 days at 25°C under aerobic conditions. During this time, the matrix is broken down using degradative enzymes such as invertases, cellulases, and amylases. This is following by adding brine (salt solution) and microorganisms like lactobacilli and *Saccharomyces roxii. A. oryzae* is prevented from growing further due to development of anaerobic conditions. Fermentation is continued for approximated 6 months and eventually results in formation of soy sauce – dark and salty liquid rich in sugars, amino acids, and vitamins.

5.3.8. Biofuel Production

As the fossil fuels have begun to diminish and petroleum costs increased, there has been a recent effort to produce fuel using resources that are renewable. Because fungi are effective degraders of lignocellulose material and high solvent producers (e.g., *S. Cerevisiae*), they have potential to be used for production of biofuel. In 1980, it was discovered that fungi are able to ferment pentose sugars along with the already-know hexose sugars. Wood and several other plant materials are composed of xylan which is rich in the pentose sugar xylose. Millions of tons of this hemicellulose materials are produced yearly worldwide and therefore, conversion of this material into fuel will be invaluable and highly beneficial. *S. Cerevisiae* can ferment xylose into ethanol when it has been pre-treated at 75°C with xylose isomerase enzyme which converts xylose into xylulose. However, other species of yeast including, *Pachysolen tannophilus*, *Candida shehatae* and *Pichia* spp can ferment xylose directly. Primarily, these species first convert xylose into xylitol (a pentitol) using the enzyme xylose reductase. This is followed by conversion into xylulose by the action of xylitol dehydrogenase. This compound is then entered into the pentose phosphate pathway. Xylose fermentation, contrary to glucose fermentation needs aerobic conditions to enhance ethanol production. This oxygen is probably required for generation of NAD^+ needed by xylitol dehydrogenase via respiration. Build of xylitol and reduction of ethanol production occurs when the conditions are anaerobic.

Several filamentous fungi have been screened for their abilities to produce ethanol. One such cellulase degrading species is *Trichoderma reesei.* It grows worldwide in soil and on dead decaying plants. It has been employed for degradation of biomass into fermentable sugars which can eventually be utilized for ethanol production. An endophytic fungus, *Gliocladium roseum,* is also in focus for this purpose. Found on trees in Patagonia, this species is able to produce several chemicals, such as heptane, benzene, and octane through degradation of cellulose. Currently, this fungi and its close relatives are being studied for their ability to produce diesel.

Even though the fungi have the potential for biofuel production, the optimization of the fermentation process and reduction of negative effects on the produced food is currently being tried.

5.4. CONCLUSION

Several foods and beverages are being produced since thousands of years by fungi. Along with this, SCP, antibiotics, enzymes, and several secretory proteins are being produced using fungi. Conclusively, fungi are being continually used on large scale by humans.

6
CHAPTER

PHARMACEUTICAL AND CHEMICAL COMMODITIES FROM FUNGI

CONTENTS

6.1. INTRODUCTION

Fungi have been highly valuable to the human economy for a long time. They have been used for yielding a wide range of commodities. Thousands of years ago, their utilization began with the production of food until the previous century when fungal biotechnology was significant in the production of several hormones, proteins, plant growth regulators, vitamins, enzymes, antibiotics, pharmaceutical drugs, and fungicides. Recently, these applications have expanded further. Fungal biotechnology and its scope, benefits, and impact are enormous and cannot be discussed and appreciated completely in this book. In this chapter, however, we will discuss the most prominent applications of fungi in the industrial production of products. Table 6.1 summarizes a list of products that are produced by fungi and have high economic worth.

Table 6.1: Economically Important Products Obtained through Fungal Cultivation

Types of products	Examples	Industrial use	Fungal Species
Enzymes	Amylase Cellulase Protease	Starch processing Fermentation application Animal feed Brewing Meat/leather Cheese manufacture	*Aspergillus niger Rhizopus oryzae.* *Trichoderma longibrachiatum* *Aspergillus oryzae Rhizopus oligosporus*
Organic Acid	Citric acid Itaconic acid Malic acid Fumaric acid	Soft drinks Chemical Beverage/food Food	*Aspergillus niger* *Candida/Rhodoturula* *Candida* *Candida*
Vitamins	Riboflavin Pyridoxine D-Erythro-ascorbic acid	Health industry Health industry Health industry	*Candida Pichia Candida*
Antibiotics	Penicillin Cephalosporin	Human/animal health Human/animal health	*Penicillium chrysogenum* *Cephalosporium acremonium*
Fatty Acids	Stearic Dicarboxylic	Food industry Chemical industry	*Cryptococcus Candida*
Alcohol	Industrial alcohol Beverage alcohol	Fuel industry Beverage industry	*Saccharomyces Saccharomyces*
Pharmaceuticals	Lovastatin Cyclosporin	Human health Human health	*Monascus rubber Tolypocladium inflatum*

Amino acids	Lysine Tryptophan Phenylalanine	Health industry Health industry Health industry	*Saccharomyces Hansenula Rhodoturula*
Recombinant proteins	Insulin Phytase Hepatitis B surface antigen	Treatment of diabetes Phosphate liberation Vaccine preparation	*Saccharomyces cerevisiae* *Aspergillus niger* *Saccharomyces cerevisiae*

(Source*: Murphy, R. A., & Horgan, K. A. (2005). Antibiotics, Enzymes, and Chemical Commodities from Fungi Fungi (pp. 113–143): John Wiley & Sons, Ltd.*)

6.2. FUNGAL METABOLISM

All fungal species are heterotrophic, i.e., they depend on other organisms for their food. These organisms can be divided into two main categories depending on the way they obtain and assimilate their nutrients. Fungi that effectively get their nutrients from other organisms are mutualistic symbionts and parasitic. Other species have the ability to decompose dead organic matter into soluble and utilizable nutrients and are called as saprotrophs. Regardless of the group they belong to, fungi have two phases of metabolism; primary and secondary. Majority of the commodities discussed in this chapter are obtained from fungi that belong to the latter group.

The primary metabolism in fungi consists of metabolites that necessary for their growth. These include proteins, carbohydrates, fats, and nucleic acids. Precursors for these primary metabolites are either obtained from the growth medium or they are synthesized by their bodies. Being essential for their growth, the primary metabolites are linked to their rapid initial growth phase. Their production increases to a maximum towards the conclusion of this phase. Further metabolism of primary metabolites may occur during the stationary phase. Some primary metabolic products formed abundantly are alcohol, enzymes, fats, and organic acids. These are produced via pathways that are common throughout the fungal kingdom and therefore, they are easy to be exploited. Screening of fungal classes for these commodities and development of techniques for their production is, therefore, more feasible. Fungal primary metabolism has been utilized extensively due to the application of recombinant DNA technologies. The primary metabolism can now be routinely exploited now for production of heterologous proteins.

Within the fungal body, secondary metabolism takes little or no part in its growth and other primary functions. When the initial log phase ends, the stationary phase begins where secondary metabolites are produced maximally. Secondary metabolites are known to be involved in the process of differentiation and sporulation and their biological roles are sometimes used economically. Primary and secondary metabolites are different in many ways. The wide range of biosynthetic origins and structures for secondary metabolites are generally not found in primary metabolites. Moreover, these metabolites are obtained from a few fungal species and can vary among differed fungal isolates within a particular species. Finally, secondary metabolites belong to a group of compounds that are closely related that can exhibit various biological properties. The primary and secondary metabolites that are obtained from various fungal species are shown in Table 6.2.

Statins, ergot alkaloids, antibiotics, and cyclosporins are some prominent examples of secondary products that have medical importance. Strobilirubin and gibberellic acid are examples of agriculturally significant fungal secondary metabolites.

Table 6.2: List of Commonly Synthesized Primary and Secondary Metabolites

Example	Fungal species
Primary metabolites	
Enzymes	*Aspergillus* sp.
Industrial Alcohol	*Saccharomyces cerevisiae*
Organic acids	*Aspergillus/Candida*
Fats	*Candida*
Polymers	*Yarrowia*
Secondary metabolites	
Antibiotics	
Penicillin	*Penicillium*
Fusidic acid	*Fusidium coccineum*
Cholesterol-lowering agents	
Lovastatin	*Monascus ruber*
Mevastatin	*Penicillium citrinum*
Immunosuppressing drugs	
Cyclosporin A	*Tolypocladium inflatum*
Plant hormones	
Gibberellic acid	*Giberella fujikuroi*

(Source: *Murphy, R. A., & Horgan, K. A. (2005). Antibiotics, Enzymes, and Chemical Commodities from Fungi Fungi (pp. 113–143): John Wiley & Sons, Ltd.*)

Several secondary metabolites are obtained from fungi which are now commercially being produced. Some of these significantly utilized

compounds will be discussed in this chapter along with the processes developed to allow their large-scale production.

6.3. ANTIBIOTICS

Antibiotics are one of the most well-characterized classes of fungal secondary metabolites. Even at very concentrations, they are able to inhibit the growth of other microorganisms. The number of antibiotics discovered is in thousands but many of them are harmful to humans, thus limiting their usage to around just 60. Clinically important antibiotics are majorly synthesized by the fungal order, Actinomycete. Various genera of fungi make antibiotics but among these, *Aspergillus*, and *Penicillium* are used for their commercial production. The largest groups of antibiotics, β-lactams are most prevalently used. These comprise of both cephalosporins and penicillins. The average market value estimated for antibiotics in 2000 was $28 billion which proves bother their medical and economic significance.

All penicillins and cephalosporins both have β-lactams rings at their cores. These rings have substitutions on their side-chains which provide different antibacterial activities to the compounds. Semi-synthetic varieties of β-lactam compounds can be synthesized artificially. Their side chains are first removed and then the β-lactam ring is used for derivation of various compounds. Structures of β-lactams ring, penicillin, and cephalosporins are shown in Figure 6.1.

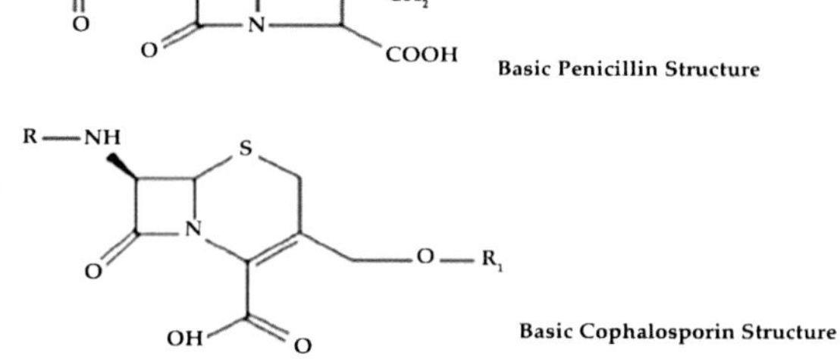

Figure 6.1: β-lactam ring and β-lactam antibiotic structures. (Source: *Murphy, R. A., & Horgan, K. A. (2005). Antibiotics, Enzymes, and*

Chemical Commodities from Fungi Fungi (pp. 113–143): John Wiley & Sons, Ltd.).

The outside of gram-positive bacteria has a layer made up of groups of carbohydrates and proteins that possess antigenic properties which elicits immune responses. This layer has polymeric peptidoglycans consisting of N-acetylglucosamine (NAG) and N-acetylmuramic acid (NAM) repeating units. Peptidoglycans are linked to penicillin-binding proteins (PBP), some of which have unknown functions. When cell wall is synthesized, it gets linked to strands of peptidoglycans through cross-linking which helps stabilize the wall. β-lactam antibiotics target this cross-linking of bacterial wall. For example, side chains of penicillins react with PBPs resulting in various effects. For example, cell lysis occurs as a result of reacting with a *transpeptidase*, PBP-1, whereas, reaction with PBP-2 causes formation of oval cells which lack the ability to replicate. Cephalosporins act similar to penicillins as they cause cell lysis by reacting and forming covalent bonds with the PBPs.

The cell wall of gram-negative bacteria has a relatively higher complexity. It usually consists of an additional membrane on the outside along with a polysaccharide-containing periplasm. Even though the peptidoglycan layer is also found in gram-negative bacteria it has PBPs but is not as thick. This renders them sensitive to β-lactam containing antibiotics.

Penicillin generally refers to a huge group of natural and synthetic antibiotics that contain a core ring, aminopenicillanic acid, with various side chain substituents. Penicillin is basically a five-membered thiazolidine ring with a side chain. When the side chains are non-polar in nature, the β-lactams are hydrophobic in nature. These include penicillin G (benzylpenicillin) and penicillin V (methylpenicillin). This category of antibiotics is produced from filamentous fungi only. Penicillin N is an example of penicillins that have polar side chains and are therefore hydrophilic in nature. These are produced by several microorganisms including actinomycetes, fungi, and bacteria.

Semi-synthetic penicillins are produced by excision of natural side chains and subsequently replaced with a different side chain yielding novel derivatives of β-lactam. Examples of such varieties are ampicillin and methicillin.

β-lactam tings in penicillins are highly reactive and therefore have high susceptibility to degradation. Penicillin can be destabilized due to their reaction with hydroxide ions which results in production of penicilloic acid which is inactive. This can also occur due to their sensitivity to acidic conditions which cause their degradation. This can be prevented by using these compounds in buffered conditions. One major limitation in the use of penicillin is that it is sensitive to penicillinases, a group of bacterial enzymes produced to help in resistance to antibiotics. β-lactamases is one of the most prevalent kind of penicillinases which inactivates the antibiotic by cleavage of β-lactam rings. In case of acylases, a different mode of action is observed where the acylamino side chain of penicillin is cleaved, making it inactive. To counteract these antibiotic-degrading enzymes, these penicillins are usually consumed with compounds like sublactams and clavulanic acid that inactivate these enzymes. Semi-synthetic penicillins are very important as they have inherent resistance to several penicillinases. For example, enzyme-resistant methicillin is often used even though it is not very effective. Interestingly, all penicillins that have resistance to β-lactamases have lesser potential compared to their parent compounds.

Soon after discovery of penicillins, cephalosporins were also found. Both of these are closely related, nevertheless, they were not immediately used as popularly as penicillins. A typical cephalosporin in cephalosporin C which is made up of β-lactam ring made up of six members with dihydrothazine unlike the thiazolidine five-membered ring present in penicillins. Majority of the synthetic cephalosporins currently in use are produced by excision of side-chain from cephalosporin C which generates 7-aminocephalosporonic acid (7-ACA). This is subsequently used for further synthesizing a variety of antibiotics. It is so cost-effective that rather than using fungal fermentations to yield antibiotics, it is more economical to use penicillin G to yield 7-ACA.

Cephalosporins are mostly less reactive compared to penicillins. Yet, certain β-lactamases attack cephalosporins more than penicillins leading to development of second, third, and fourth generation cephalosporins which have various antimicrobial properties, absorptions, side-effects, metabolism, and susceptibility to resistance by microbes. Cephalothin and cephazolin are first-generation cephalosporins. Cefamandole and cefaclor are examples of second-generation cephalosporins. Cefotaxime and cefixime are third-generation cephalosporins and cefepime is an example of fourth-generation cephalosporins (Figure 6.2).

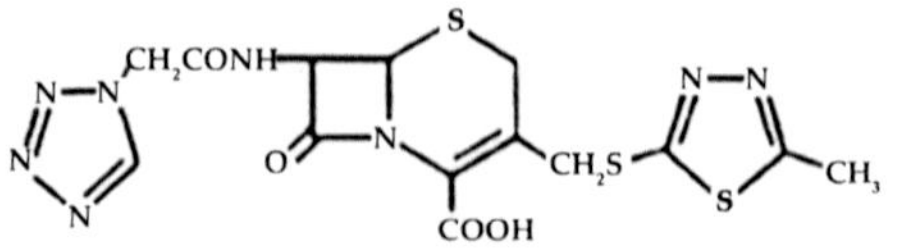

Cefazolin (1st generation)

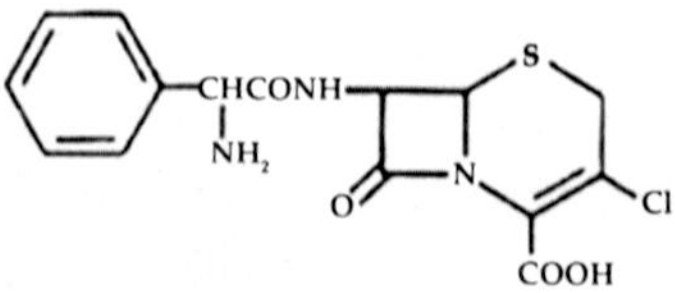

Cefachlor (2nd generation)

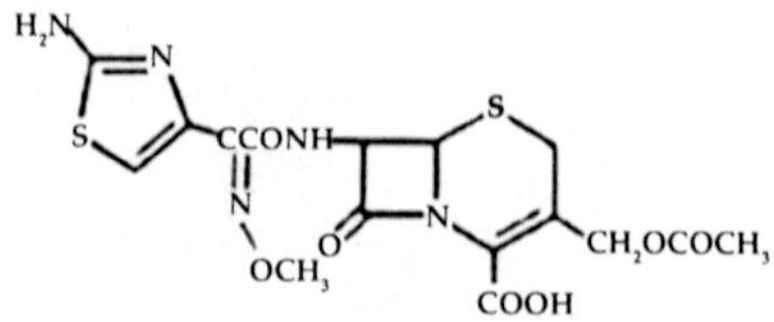

Cefotaxime (3rd generation)

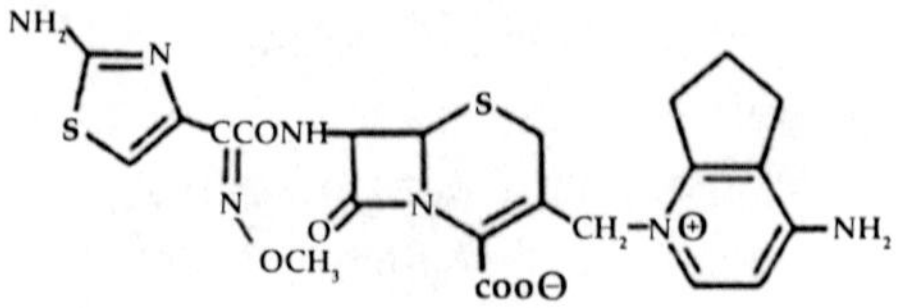

Cefpirome (4th generation)

Figure 6.2: Structures of cephalosporins. (Source: *Murphy, R. A., & Horgan, K. A. (2005). Antibiotics, Enzymes, and Chemical Commodities from Fungi Fungi (pp. 113–143): John Wiley & Sons, Ltd.*).

6.3.1. Cycles of Antibiotic Production

Both penicillins and cephalosporins are produced by pathways that have the first two common steps. Three amino acids, L-cysteine, L-α-aminoadipic acid and L-valine react to form a tripeptide called as ACV. Subsequently, this compound is acted upon by IPN synthase to produce isopenicillin N (IPN). This intermediate is common for production of both penicillins and cephalosporins. Hydrolysis of IPN produces 6-amino penicillanic acid (6-APA), which is then converted into various types of penicillins. On the other hand, IPN can be converted into penicillin N by epimerization. This compound, with the help of enzymes, gives rise to deoxycephalosporin C which can be modified to either form cephalosporin C or cephamycin C.

For semi-synthetic penicillins, 6-APA reacts with different types of organic acids which give rise to different types of penicillins. Alternatively, 6-APA is produced by removing the side chain of penicillin G. The removal of side chain from cephalosporin C results in formation of 7-amino cephalosporonic acid (7-ACA). Several cephalosporins can be produced by reacting 7-APA with various organic acids.

Antibiotics that do not have β-lactams are also produced by fungi. These include Griseofulvin (originating from *Penicillium griseofulvin)*. It prevents fungal growth by interfering with microtubule assembly which is part of process of mitosis. Another example is that of fusidic acid, a steroidal compound synthesized from *Fusidium coccineum*. Gram-positive bacteria, including the bacterial strains that are resistant to β-lactams, are killed by this antibiotic.

6.3.2. Antibiotic Production Industrially

Industrial production of antibiotics through fermentation is inefficient. 10% of the provided carbon sources is incorporated in the synthesis of antibiotics. When limited amounts of nitrogen, carbon, and phosphorus are added in the fermentation medium the fungi grow at a slow rate which helps in optimal formation of antibiotics. Several procedures are employed for antibiotic production, which have the same basic process. A preserved culture stock is used to inoculate the primary culture. Culture stock can be spores that were either lyophilized or cryopreserved. Several others methods are used which are not covered here.

Agar slants or liquid cultures are used for carrying out primary cultures which is subsequently used for inoculation of secondary cultures. These cultures, maintained on particulate matter or on agar-coated bottles, eventually generate large amounts of spores.

A suspension is formed from the spores produced in secondary cultures. This is used for inoculation of liquid media. The antibiotic-producing fermentation is carried out under strict hygiene and aseptic conditions to avoid contaminations. The industrially used fungal strains are exposed to extensive selections pressures and mutations during isolation. They are therefore less robust compared to wild-type strains.

The scaling up of the process can be done in three to four stages depending of the size of the process. Generally, secondary spore suspension produces a seed culture with a volume less than 10 L. After allowing growing for a certain period of time, this culture is used for inoculation of 20,000 L medium. Eventually, this medium will be used as inoculum for 300000 L culture which is the final stage culture. The production cycle largely depends on product's nature. Since antibiotics are products of secondary metabolism, it is necessary to ensure that the fungi grow in a limited way so that they stay in secondary metabolism phase. This is achieved by adding certain nutrients in limited amounts at the right time so that metabolic phase changes at the right time which is the key to antibiotic production. For penicillin production industrially, the fermentation medium has a limited supply of glucose.

In the downstream process, it is important to purify the antibiotics from the medium, mycelia of fungi, and other metabolites. This process varies depending on the antibiotic type and the employed process. Usually, the fungal mass is removed using filtration or centrifugation. The antibiotics are further purified using processes like ultrafiltration, solvent extraction, chromatography, and drying in order to prepare them for pharmaceutical use.

6.4. PHARMACOLOGICALLY ACTIVE PRODUCTS

Other than antibiotics, fungi produce a variety of secondary metabolites that have medical significance. These include cyclosporine A and statins (compounds with ability to lower cholesterol levels). In this section, we will discuss some other secondary metabolites including gibberellins and alkaloids.

6.4.1. Cyclosporin A

Organ rejection has been reduced greatly since the use of immunosuppressive drugs became clinically applied. Cyclosporin A, produced through fungal biotechnology, acts as a commonly used immunosuppressant. Chemical synthesis of this drug has show how complicated secondary metabolism in fungi is. *Tolypocladiuminflatum* is the fungal isolate which produced cyclosporin A. It was first discovered through screening of soil sample in Norwegian. Antigen introduction as result of organ transplant induces immune responses by triggering production of T-lymphocytes. This drug functions by inhibiting T-lymphocytes' production of interleukin-2, thus preventing an immune response. Interleukin-2 also plays a role in diseases like eczema and psoriasis where it causes inflammation responses. Treatment of these conditions is also done through employment of this compound.

Cyclosporine A is composed of a highly methylated cyclic peptide. Like other secondary metabolites, cyclosporine A also has several analogs formed from *T. inflatum*. Out of a variety of 25 cyclosporins, 17 exhibit antifungal properties and 2 acts as immunosuppressants. Optimization of fermentation medium has resulted in large quantities of this compounds (in the ranges of grams per liter). This has also been possible through mutations introduced for improving strain characteristics. Despite efforts, economic production of cyclosporin A is currently feasible only through natural fermentation and not through chemical synthesis.

6.4.2. Statins

Statins are a group of compounds that are known to interfere with synthesis of cholesterol. They do so by inhibiting an important enzyme in the process, 3-HMG-CoA-reductase. Through their acidic properties, these organic acids are able to influence the enzyme activity resulting in reduced blood cholesterol levels. Mevinic acids are the most commonly and commercially used statins. Under this category comes lovastatin (originated from *Monascus ruber*) and mevastatin (originated from *Penicillium citrinum*). Chemical or microbial transformation can be employed for conversion of mevastatin and lovastatin into ML-236A and monacolin J, respectively. These compounds have different effectiveness due to their varied affinities for 3-HMG-CoA-reductase. Several researches have shown that many fungal species are able to produce a variety of compounds that have the power to lower cholesterol levels. Interestingly, these products can be synthesized commercially via liquid fermentation (e.g., *Aspergillus terreus* produces lovastatin) or solid-state fermentation (e.g., *Monascus rubber* produces lovastatin).

6.4.3. Alkaloids

One of the grains pathogenic genus, *Claviceps* is able to produce a group of secondary metabolites called as alkaloids. These fungi synthesize alkaloids in their sclerotium, a structure formed which helps in survive during winter. These compounds are medically important and function through various mechanism. Their particular target is the central nervous system due to which they cause convulsions and hallucinations. Within the core of their structure in an alkaloid ring which tetracyclic and the nitrogen at 6th position has methyl group attached to it. Tryptophan and mevalonic acid are used for derivation of this core structure. A variety of alkaloid compounds varying in their potencies, toxicities, and complexities are formed by its modifications. Furthermore, these derivatives are used for chemically synthesizing several other pharmacologically active compounds. One example is that of lysergic acids, released *by Claviceps paspali*, which are synthetically converted into lysergic acid diethylamide (LSD). These alkaloids also have medical significance due to their negative effects on human health. During the eighteenth century, for instance, bread contaminated with alkaloids was responsible for the wide-spread ergotism.

However, several alkaloids have now been isolated due to their usefulness. Ergometrine and methylergometrine are two such examples. These compounds are known to induce uterine smooth muscle contractions and to prevent excess blood loss after childbirth by acting as a vasoconstrictor.

6.4.4. Gibberellins

This group of compounds consists of 19 or 20 atoms of carbon which form diterpenoid compounds. They are known to regulate several developmental processes in plants. Some examples of their effects are; stimulation of bolting, secretion of hydrolytic enzymes, and induction of cell division and elongation. *Gibberella fujikora* are the major producers of gibberellins which use mevalonic acid as the precursor. Other fungal species have also been identified to be producers of these secondary metabolites including *Neurospora crassa*, *Sphaceloma manihoticola*, *Azospirillium lipoferrum* and *Rhizobium phaseoli. Industrially,* they are exploited for ripening of fruits, improvement of fruit quality, and malting. The chief organism that secretes these compounds is *G. fujikora.,*Some mutant strains have been produced which are able to produce gibberellins at quantities as large as grams per liter.

6.4.5. Endophytic Compounds

There is a constant need for development of new antibiotics and chemotherapeutic agents which are effective, less toxic, and environmentally friendly. This is because the infectious microorganisms often develop resistance to the existing antibiotics. Even though both pharmaceutics and agriculture industries continuously focus on finding novel products for this purpose, the natural selection has played a more significant role compared to the synthetic mechanisms to help discover new natural products. Because these natural products have some natural function, it is likely that they will be secreted by organisms that are present in novel ecosystems.

Interestingly, recent work has revealed that plants can act as reservoirs for several other organisms called as endophytes. Fungi that belong to this category come under a group that has many phyla. They are ascomycetes that are very diverse in nature. They are found on a number of plant species including ferns, lycophytes, mosses, hornworts, liverworts, equisetopsids, and seed plants. Particularly, they inhabit the plant tissues without having negative effects immediately because they undergo a dormant stage as soon as they are inside the host plant. This dormancy may continue through the lifetime of the plant or it may end when the conditions are favorable for fungi to become pathogenic. These include alterations of environmental conditions or change in the oncogenic state of the plant. The host plants allow this colonization as it may be helping them to adapt to certain stressful factors, both biotic and abiotic. Interestingly, tolerance of the host plant to biotic stress has been found to improve in the presence of natural products originated from fungi.

Endophytic fungi are not well understood yet but they are expected to yield various products that may have potential to be beneficial to humans. This is evident from Table 6.3 which lists a variety of fungal endophytic species and their products. Several other fungal endophytic characteristics makes them suitable for manipulation at industrial level. Because there hasn't been much focus in screening and harvesting these organisms and their products industrially, their vast resource remains unknown and undiscovered.

Natural fungal products that are produced by endophytes have various bioactivities. They act as antibacterial, anti-cancerous, anti-viral, immunosuppressive, and insecticidal agents. The fungus *Taxomyces andreanae* that grows on the plant *Taxus brevifolia* produces paclitaxel (Taxol (Figure 6.3)). This directed focus on studying the *Taxus* species along with other plants to screen for endophytes that produce Taxol. This product was first isolated from the plants' roots, branches, and barks. It is basically a tetracyclic diterpene lactam which has anti-cancerous properties against lung, breast, head, neck, and ovarian cancers. *Taxus* plant species are on the verge of extinction and the product is produced in their barks at very low concentrations (0.01–0.05%). Unfortunately, current methods from extraction of taxol from tree barks are very costly and not very efficient.

Table 6.3: Endophytic Fungal Species, their Hosts, and their Bioactive Products

Endophyte	Host plant species	Metabolic product	Activity
Colletotrichum gloeosporiodes	*Artemisia mongolica*	Colletotric acid	Anti-microbial
Muscodor roseus	*Erythophelum chlorosachys*	Volatile antibiotic	Anti-microbial
Muscodor albus	*Cinnamomum zeylanicum*	1-Butanol, 3-methyl acetate	Anti-microbial
Phoma sp.	*Taxus wallachina*	2-Hydroxy-6-methyl benzoic acid	Anti-microbial

Fusarium sp.	*Selaginella pallescens*	CR377	Antifungal
Cryptosporiopsis quercina	*Tripterigeum wilfordii*	Cryptocandin	Antifungal
Nodulisporium sp.	*Bontia daphnoides L*	Nodulisporic acids	Anti-insecticidal
Semiatoantlerium tepuiense	*Maguireothamnus speiosus*	Taxol	Anti-cancer
Taxomyces andreanae	*Taxus brevifolia Nutt*	Taxol	Anti-cancer
Tubercularia sp.	*Taxus mairie*	Taxol	Anti-cancer
Pestalotiopsis microspore	*Taxus wallachina*	Taxol	Anti-cancer
Sporomia minima	*Taxus wallachina*	Paclitaxel	Anti-cancer
Rhinocladiella sp.	*Tripterygium wilfordii*	22-Oxa-[12]-cyto-chalasin	Anti-cancer
Pestalotiopsis guepinii	*Wollemia nobilis*	Taxol	Anti-tumor
Fusarium subglutinans	*Tripterygium wilfordii*	Subglutinols A & B	Anti-tumor

(Source: *Tejesvi, M.V., Nalini, M.S., Mahesh, B. et al. (2007) New hopes from endophytic fungal secondary metabolites. Boletın de la Sociedad Quımica de Mexico, 1(1), 19–26*)

A new approach for producing Taxol is screening and fermentation of endophytic fungi that produce Taxol. There has been a quick progress recently in developing and utilizing these fungi for Taxol production. At industrial level, it is necessary to optimize the conditions of fermentation in order to maximize its yield. This approach will be cost-effective and it will also protect the *Taxus* species from becoming extinct. Taxol production in this way is expected to offer numerous industrial and social benefits.

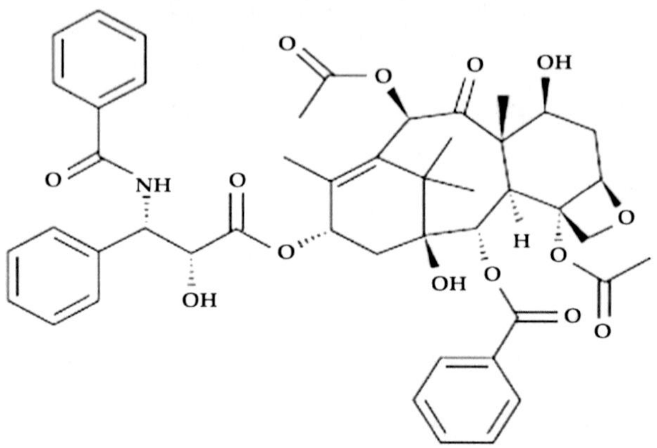

Figure 6.3: Taxol's chemical structure. (Source: *Murphy, R. A., & Horgan, K. A. (2005). Antibiotics, Enzymes, and Chemical Commodities from Fungi Fungi (pp. 113–143): John Wiley & Sons, Ltd.).*

Other than taxol, natural products originating from other endophytic microbes have also been isolated and found to prevent or even destroy growth of dangerous agents that cause diseases, such as fungi, bacteria, viruses, phytopathogens, and protozoans that are pathogenic to plants, animals, and humans. Another endophytic fungal species is *Cryptosporiopsis quercina*, which grows on the medicinally important hardwood plant *Tripterigeum wilfordii*, native to Eurasia. This fungus is shown to have inhibitory effects *in vitro* on *Trichophyton* sp. and *Candida albicans*, fungal species that are pathogenic to humans.

Fungi are increasingly becoming problematic to human health as they often cause infections in AIDs and other immunocompromised patients. It is necessary to look for anti-mycotic agents that fight against these fungal species. *C.quercina* produces cryptocandin, an anti-mycotic peptide which has been isolated and characterized. Its structure consists of various amino acids that are hydroxylated including one novel amino acid. This compound is found to be related to other well-known anti-mycotic compounds like pneumocandins and echinocandins. Interestingly, this compound also has activity against *Sclerotinia sclerotiorum* and *Botrytis cinerea*, which are fungal plant pathogen. Currently, this compound is being studied and tested for its effectiveness against skin and nail diseases.

Some natural products from endophytes are also able to affect animal immune system. One such example is of *Fusarium subglutinans* that grow on *T. wilfordii.* These endophytes produceSubglutinols A and B which have been assessed using lymphocyte reaction assay and both are shown to have IC50 values of 0.1μM. In the same study, cyclosporins were also shown to have the same potential as subglutinols. Currently, their function in the endophytes is not known. But these natural products are being studied further for their application as immunosuppressive agents.

Recently, endophytes have proved to be a reservoir of several bioactive products that are not yet studied but are very diverse in their chemical properties. They are likely to have evolved and produced as a result of multiple factors including ecological, evolutionary, and environmental.may acute and chronic conditions is humans are expected to be treated through drugs developments from endophytic fungal species. Because only a small number of studied endophytes has yield such a vast range of bioactive compounds, it is clear that the chances of discovering more novel, potent, and reliable bioactive products in the future are enormous.

6.5. CHEMICAL COMMODITIES

Molds and yields are often industrially utilized in production of a variety of chemicals. The most important worldwide among these is citric acid.

6.5.1. Citric Acid

In citrus fruit, the chief organic acid found is citric acid. There is an increasing demand for this chemical which is met through fungal fermentation systems of *Aspergillus niger* and the genus *Candida* in the presence of feedstock rich in carbohydrates. Initially, commercial production of citric acid was done through surface fermentation of *A. niger*. Once techniques developed for submerged fermentation systems, it was groundbreaking for production of citric acid. Citric acid is used popularly in soft drinks and confectionary products. Recently, it has been used for reduction of oxidation on metals and for cleaning metal surfaces as it acts as an agent for complexing metals. To meet the increasing demands for citric acid worldwide, it is increasingly being produced through fermentation at an annual rate of 500000t.

Even though several species of fungi and yeasts have been employed for its production, fermentation of *A. niger* offers several advantages which is now the prevalently used species. It is easy to handle, produces high yields and is able to ferment raw material that are cheap, such as molasses, hydrocaarnons and starch materials. If available and cost-effective, purified glucose syrup, cane molasses, and sucrose can also be used. Among these, molasses are considered to be the most cost-effective and contain up to 50–55% sugar content (in the form of fructose, glucose, and sucrose). Conditions that are employed for *A. niger* fermentation for the production of citric acid varies a lot in the literature. The basic nutrients that must be present in the fermentation medium are sources for nitrogen, carbon, phosphorus, and some micronutrients. The fermentation process is highly affected by the pH, temperature, and aeration.

A. niger utilizes different carbon sources variably. For instance, it is able to readily take up glucose and fructose, the simple sugars. At the industrial level, the preferred sugar is sucrose. The mycelium of the fungus releases an invertase enzyme that hydrolyzes sucrose into its monomeric sugar under acidic fermentation conditions. Maximum yield is obtained from *A. niger* fermentation when supplied with sugar concentrations of 14–22 %. When sugars are present at lower amounts, the yield of citric acid also lowers and there is build-up of oxalic acid.

The preferable sources of nitrogen for *A. niger* fermentation is ammonium sulfate, sodium nitrate, potassium nitrate, ammonium nitrate, and urea. Amount of phosphorus used in the medium also influences the citric acid production. High levels of this element lower the production of acids even though it induces growth of the fungal cells. The most suitable source of phosphorus is Potassium dihydrogen phosphate (0.1 %). The optimum pH that has to maintained for citric acid production in below 2. Because this process is aerobic, the yields can be enhanced and fermentation times reduced when the medium is aerated at higher rates.

Several trace elements also regulate this citric acid yields that are obtained. When they are supplied in limited amounts, the growth of fungi becomes limited and accumulation of citric acid increases significantly.

6.5.2. Filamentous Fungi as Producers

Several types of systems are used for citric acid production. These are discussed in the following sections.

6.5.2.1. Surface Fermentation

Large-scale production of citric acid was for the first time done through surface culturing. Even though more advanced methods like submerged fermentation are now in use, surface fermentation is quite popular for its ease to be set up, install and operate and is cost-effective. Mycelia are allowed to grow spread on 50–100 L trays of stainless steel or aluminum which are stacked in aseptic chambers.

The source of carbohydrate is used at a concentration of 15% and the pH and temperatures employed are in the range of 5 to 7 and 28 to 30°C, respectively. After media preparation and sterilization, it is cooled and flown onto trays' surfaces. It is then inoculated with spores, either in the form of suspension or blown with air. The spores germinate and develop into mycelia under 40–60% humidity.

Considerable amounts of heat are dissipated during the fermentation process, therefore, it is necessary to aerate the media. This aeration maintains humidity, temperature, and oxygen levels within the fermentation media. During the course of fermentation, the pH lowers to a value lower than 2.0. If this value goes above 3.0, it will lead to formation of oxalic acid and gluconic acid. After fermentation for 8 to 12 days, the liquid is removed from the trays and the mycelium are collected and processed further. This process has a yield of around 70–75%.

6.5.2.2. Submerged Fermentation

Commercial production of citric acid is often carried out using this system. It offers the advantage of occupying lesser space and labor and increased productivity. It is carried out in a tower fermentor or a stirred tank reactor.

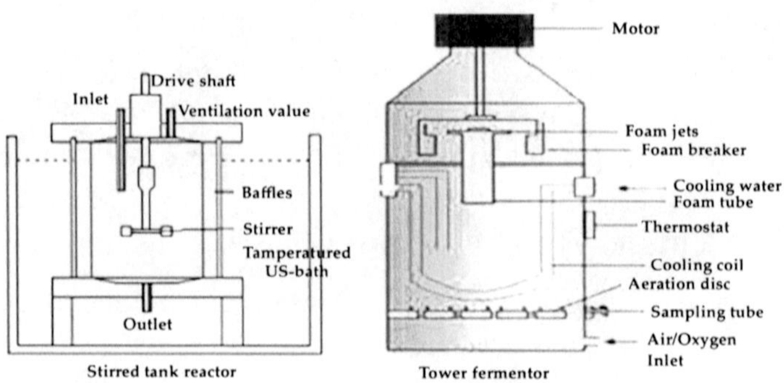

Figure 6.4: Reactors used for production of citric acid. (Source: *Murphy, R. A., & Horgan, K. A. (2005). Antibiotics, Enzymes, and Chemical Commodities from Fungi Fungi (pp. 113–143): John Wiley & Sons, Ltd.*).

The bioreactors that are used are resistant to acid, it is particularly important because of development of acidic pH as a result of fermentation and due to corrosiveness caused by the produced citric acid (Figure 6.4). These fermenters are designed in a way to allow proper aeration by sparging air from the base.

In submerged fermentations, media is prepared by specific dilution of carbon source, followed by introduction of other nutrients and sterilization in line or within the fermentor. This setup can be inoculated by using spore suspension or precultured mycelia. Usually, spore inoculation is added simultaneously with a surfactant to allow thorough dispersion. 10% of the primary culture can be used as an inoculum in case of precultured mycelia. During the fermentation process, 0.5–1.5 VVM (VVM (volume of air per volume of culture medium per minute) is the rate at which air is sparged within the medium. This type of fermentation can be done in a continuous manner or in the fed-batch manner, the latter being more frequently employed. When the conditions are ideal, it takes 5 to 10 days for completion of fermentation in this system.

6.5.2.3. Fermentation in Solid State

This process developed in Japan is also referred as the Koji process. It is a relatively simple process which is simple to surface fermentation. Wheat brans, fruit wastes, rice, and sweet potato fibrous residues are used as raw materials for this process. The source of carbohydrate is moisturized up to 70% with water, exposed to steam to allow sterilization, and finally spread on tray surface. The conidia produced by *A. niger* are used for inoculating the media. At a fermentation pH of 5.5, fungal amylase hydrolyzes the starch and the products are subsequently changed into citric acid. This type of fermentation process completes in around 4 to 5 days. Drawback of this method is accumulation of trace elements which are difficult to remove through traditional procedures.

6.5.3. Production by Yeast

Commercial production of citric can also be carried by employing yeasts. The frequently used strains include *Candida guilliermondii, Saccharomyces lipolytica, Candida olephila, Candida citroformans, Hansenula anomala* and *Candida tropicalis*. Using yeasts offers several advantages over filamentous fungi. For example, yeasts are able to tolerate elevate levels of sugar, they are not sensitive to trace elements thus can be grown in untreated crude sources of carbon, they can be cultured using continuous system and their rate of fermentation is high. Commercially yeasts are used for citric acid production in tower fermenters which contain efficient cooling systems. For inoculum preparation, another small fermentor from where a particular amount of media transferred into the main fermentor. Depending on the fungal strain, the optimum temperature and pH are in the ranges of 25–37°C and greater than 5.5, respectively. Employment of *Candida* for production of citric acid is done using cane molasses as the carbon source in a continuous system. In addition, a process which uses *Candida* strains with n-paraffin as the carbon source has also been developed. Because the petroleum products are costly, this process is uneconomic and is not yet applicable for large-scale production of citric acid.

6.5.3.1. Metabolic Pathways for Citric Acid

The metabolic pathway used for citric acid production is unclear but is thought to be an incomplete version of Kreb's Cycle. The glucose is first metabolized to elevated cellular levels of oxaloacetate. This, in turn, increases the citrate synthetase activity and simultaneously decreases its breakdown by α-ketoglutarate dehydrogenase. Citrate is synthesized by condensation reaction between acetyl-CoA and oxaloacetic acid in the presence of citrate synthetase (Figure 6.5). This enzyme is allosterically regulated. To increase citric acid yield, more oxaloacetic acid is produced through Kreb's cycle and through anaplerotic reactions. The elevated cellular levels of citric acid necessitate it to be excreted out of the cells.

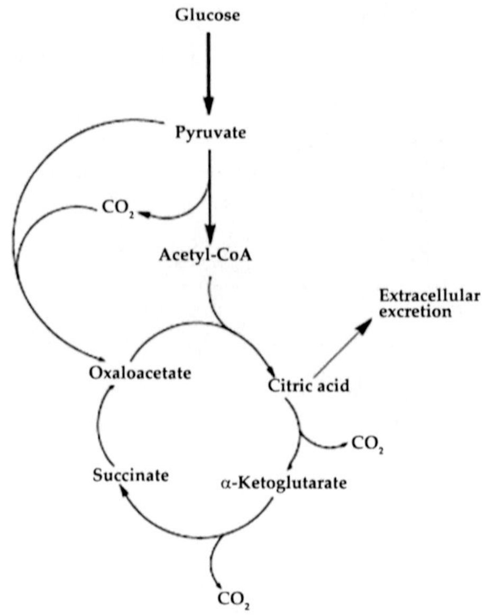

Figure 6.5: Metabolic pathway for synthesis of citric acid. (Source: *Murphy, R. A., & Horgan, K. A. (2005). Antibiotics, Enzymes, and Chemical Commodities from Fungi Fungi (pp. 113–143): John Wiley & Sons, Ltd.*).

6.5.3.2. Recovery of Citric Acid

Once fermentation is carried out for 5 to 10 days, the fungal cells are removed from the liquid media through filtration or centrifugation. Yeast fermentors that are neutralized using calcium carbonate first require to be acidified using mineral acid before filtration. When lime is added to the filtrate or supernatant, citric acid is precipitated out in the form of calcium citrate tetrahydrate. Calcium salt is filtered, washed, and then mixed with sulfuric acid in an acidulator. This results in precipitation of calcium sulfate which is eventually filtered. The rest of the solution containing citric acid is reacted with active carbon, poured into cation exchanger, and eventually crystallized by evaporation. Another way to obtain citric acid is by using tributyl phosphate or long-chain secondary or tertiary amines. At low temperatures, citric acid is mixed with solvent and subsequently, at higher temperatures, it is re-extracted into water. Finally, the purified citric acid is concentrated and crystallized.

6.5.4. Itaconic Acid Production

This chemical is used in polymer synthesis for emulsion paints and to change specific features of vinyl polymers. *Aspergillus itaconicus* was the first fungal strain that was identified to be a producer of itaconic acid. Currently, A. terreus and its mutants are now prevalently used for this production. Commercially, this organic compound is produced in the presence of carbon sources glucose and inorganic salts, molasses or beet molasses. The presence of calcium and zinc is necessary in the growth medium. After preparation and sterilization of media, it is inoculated by spore suspension or pre-cultured mycelium. The most commonly and preferably used system is the submerged fermentation even though fermentation is sometimes carried on trays. Tanks made up of stainless steel are used which have a system of allowing aeration, agitation, and cooling. Itaconic production occurs at higher fermentation temperatures of up to 40°C and the initial pH has to be around 2. Once initiated and itaconic acid accumulated, a neutral pH is more favorable for further production of itaconic acid. To enhance productivity, the fermentation medium is continuously aerated to provide aerobic conditions. Depending on the carbon source employed, 72 hours of fermentation can yield up to 60% yield of this compound.

Cells of *A. itaconicus* are able to metabolize the carbohydrate into pyruvate through glycolysis. Subsequently, this is changed into aconitic acid via citric acid cycle. Finally, aconitic acid is converted into itaconic acid with the help of an oxygen-dependent enzyme, aconitic acid decarboxylase.

Similar to other products, itaconic acid is harvested from the fermentation medium through filtration of mycelium and the filtrate is then evaporated and crystallized or it is obtained by solvent extraction of ion exchange.

6.5.5. Production of Vitamins

Vitamins are important nutrients that are needed in minute amounts. The required values in humans and animals body have been well defined. Because there are unbalanced diets, food processing, shortage of food and spread of diseases, there is a vast need for additional vitamins other than the vitamins that are present in plant and animal sources. These additional vitamins are synthesized by chemical or biotechnological processes through bioconversion and fermentation methods (Table 6.4). Organic chemical production of vitamins is carried for multiple types of vitamins. Nevertheless, production of these compounds often depends on processes involving microbiology. Vitamins C, B2 (riboflavin), D2 (ergosterol, B12 (cyanocobalamin), B13 (orotic acid) and F group are all produced through fermentation methods.

Table 6.4: Vitamins Produced Industrially

Vitamin	**Organic extraction**	**Chemical synthesis**	**Microbial synthesis**			
			Bacteria	Fungi	Algae	Global production per year (tons)
Vitamin B1 +	+					2000
Vitamin B2 +	+		+	+		2000
Niacin (B3,PP) +	+		+			8500
Pantothenic acid (B5) +	+					4000
Vitamin B6 +	+					1600
Biotin (B8) +	+					3
Folic acid (B9) +	+		+			300
Vitamin B12	+					10
Vitamin B13	+					100
Vitamin C +			+	+		70000
Vitamin A +			+			2500
Provitamin D2	+		+	+		
Provitamin D3 +	+			+		25
Vitamin E +		+				6800
Vitamin F	+	+			+	1000
Vitamin K2 +	+				+	2
	+	+				

(Source: *Murphy, R. A., & Horgan, K. A. (2005). Antibiotics, Enzymes, and Chemical Commodities from Fungi Fungi (pp. 113–143): John Wiley & Sons, Ltd.*)

6.5.5.1. Riboflavin (Vitamin B2)

Human nutrition and animal feeds often require riboflavin. It is produced by synthetic and more prevalently, by fermentation methods. Initially, the application of fermentation systems was not very applicable as it was a lot more expensive compared to chemical processes. Later, developments in fermented organisms led to increased feasibility of fermentation systems which are now popularly used in production of riboflavin. The bacteria (*Clostridium* sp.), yeasts (*Candida* sp.) the ascomycete fungi *Ashbya gossypii* and *Eremothecium ashbyi* are all good producers for riboflavin. Due to its genetic stability, *A. gossypii* is more preferably used species.

This fungal species is fermentation in a medium containing soya bean mean and soya bean oil. Within the growth medium, when the fungi use up all the glucose and the fungal cells undergo the later phases of growth, it starts to produce riboflavin. During the early growth phase, the present glucose in changed into droplets of lipids which are later utilized in production of riboflavin. This production is stimulated in the presence of ribitol or glycine which acts as precursors for the pathway for synthesis of riboflavin.

Alternatively, the yeast species, Candida famata can also be fermented to commercially produce riboflavin. This strain has also been mutated to enhance its riboflavin productivity to up to 200g/L after allowing fermentation for 8 days.

6.5.5.2. Vitamin D

Large amounts of fortified foods and animal feeds contain vitamin D_2 and D_3. In addition, these compounds are also used for anti-rachitic treatments. When the provitamin D_2 (ergosterol) present in yeast is exposed to UV radiation, it gets converted into Vitamin D_2 (ergocalciferol) (Figure 6.6). Ergosterol can be produced in high amounts through developed fermentation systems using *Saccharomyces cerevisiae*. This species is able to synthesize a variety of sterols including 90% ergosterol, ergosta-5, 7, 22, 24 (28)-tetraen-3β-ol, zymosterol, and lanosterol. Yeast cells are able to carry out both aerobic and anaerobic metabolism for utilization of carbohydrates. Depending on the levels of carbohydrates and oxygen that are present, the yeast cell determines which process it will undergo. Fermentation process for ergosterol production is carried out using fed-batch methods. This helps prevent inhibition that occurs as a result of lack of nutrients and an oversupply of carbohydrates. The growth of yeast cells is promoted in the presence of glucose and when it gets used up, the growth of yeast cells becomes limited. As a result, the cells adapt to the conditions and continue their growth by consumption of ethanol as the carbohydrate source which was liberated during the first growth phases. Overall, the process occurs in two phases, during the first phase, content of ergosterol increases followed by the second phase where the rate of growth becomes limited.

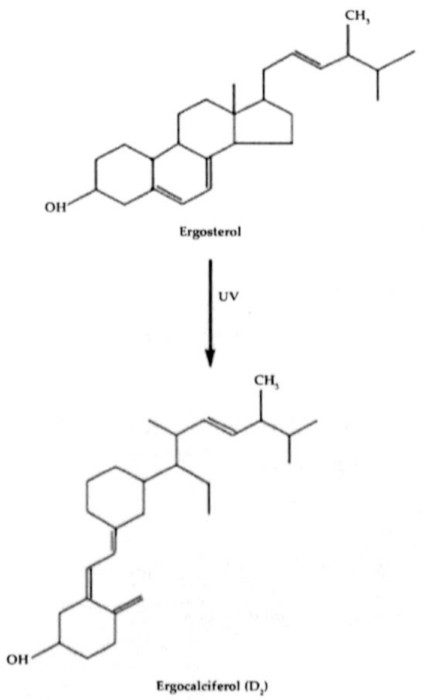

Figure 6.6: Biosynthesis of Vitamin D. (Source: *Murphy, R. A., & Horgan, K. A. (2005). Antibiotics, Enzymes, and Chemical Commodities from Fungi Fungi (pp. 113–143): John Wiley & Sons, Ltd.*).

6.5.6. Pigments in Fungi

Production processes involving fungi cause lesser pollution compared to chemical processes. For example, stains, and dyes using in textile and food industries are now popularly produced using fungal pigments rather than the environmentally-less friendly chemical processes for their synthesis and disposal. The latter processes have adverse effects on the environment. The prospects and global markets for usage of pigments that are carotenoid fungal polyketide pigments has been strengthened since the carotenoids have been approved for use as food dyes. Carotenoid utilization as food colorants offers several advantages; it environment-friendly, safe, independent of seasonal availability of raw material, and lowers the strains variability from batch to batch. Interestingly, pigments obtained from fungi are diverse in their chemical compositions, structures, and color ranges.

β-carotene is an example of carotenoid that is used for a range of foods and is produce by several fungal species including *Mucorales*. The species *Phaffia rhodozyma* produces the carotenoid astaxanthin. This is the same pigment which is found in boiled shells of crustaceans and in flesh of salmonids. The farmed salmonids are fed with this fungal species and as a result, their white muscles attain a reddish color normally found in wild fish. This product has high economic value.

Filamentous fungi are a source of polyketide pigments including quinones (naphthoquinones and anthraquinones), flavins (riboflavin) and dihydroxy naphthalene melanin. *Eurotium* spp., *Fusarium* spp., *Curvularia lunata* and *Drechslera* spp. produce anthraquinones which are yellow-colored pigments. This category includes catenarian, chrysophanol, cynodontin, helminthosporin, tritisporin, and erythroglaucin which are used as dyes. Some other yellow pigments like falconensones, falconensins, and epurpurins are produced by *Emericella* spp.

Rubropunctatin and monascorubrin are two examples of insoluble orange pigments that are produced by Monascus species. When these pigments are reacted with amino acids, they are converted into soluble red pigments – rubropunctamine and monascorubramine. These pigments are used as food colorants and food dyes. When the *Monascus* species are fermented in the presence of amino acids, they synthesize variants of pigments that have enhanced functional properties in the ranges of colors to violet to red and orange to red. One drawback of fermenting Monascus is that the rice substrate on which it is fermented has a toxic substance, mycotoxin citrinin. This makes this species unsuitable for commercial production of food dyes.

Conclusively, fungi have a high potential for being a source of food dyes and food colorants which can be industrially used.

6.6. YEAST EXTRACT

Since the existence of mankind, there has been a constant challenge to maintain the increasing demands of human supply to meet the needs of increasing populations. Every culture has long been using microbes for transforming or producing food. Among these microbes, fungi have played a significant role in food processing, improvements in texture, nutritional value, appearance, and flavor of the food. The first fungal utilization in food industry was done in Germany. In soups and sausages, Torula yeast was cultured and added for improving nutritional value. Autolysis along with other techniques is used for completely or partially solubilizing yeast cells. The slurries can be further processed and converted into several products useful for laboratories, food industries, animal feeds, and fermentation media. The main products that are formed include yeast cell components such as cells walls and proteins released by mechanical damage to the cells, soluble components in the form of grains, powder, paste or liquid, concentrates of β-galactosidase, and yeast invertase. Commercially most significant products are; yeast extracts, yeast hydrolysate, and autolyzed yeast extract.

Studies were done on spent yeast from brewery industry during the last century. This resulted in the utilization of yeast extract in the industry of food. This product has been approved by FDA to be used as natural food flavorings. They are added into products like cheese spreads, vegetable products, bakery products, meat products, soups, gravies, sauces, seasonings, and seafoods. Nutritional foods specially formulated for health, baby feeds, supplementations for growth enrichment growth media for microorganisms and other biological culture systems all contain yeast extract as this product is a rich source of peptides, minerals, vitamins, and amino acids.

There are basically three types of processes used for manufacture of yeast extract; autolysis, plasmolysis, and hydrolysis. In autolysis, the yeast cells are stimulated to degrade themselves and release their cellular components. This is done by activating certain inherent cellular enzymes by setting appropriate conditions of temperate, pH, and time in the presence of additional enzymes and reagents. The extent of hydrolysis can be estimated by the amount of free amino acids that are obtained from the yeast extract. The flavor of food is directly or indirectly influenced by the presence of these free amino acids.

The process prevalently used in Europe is plasmolysis (Figure 6.7). This process is carried out by exposing the cells to high salt concentrations which results in excessive loss of water and removal of cytoplasm from the cell wall. As result, cell death and the degradation of cell occurs. This process is feasible because it does not require any special equipment. It only needs salt which is quite cheap and easily available. However, the yeast extract obtained has a high salt concentration, which is a drawback of this method. The hydrolytic process is carried out by the help of hydrochloric acid at specific pressure and temperatures. This results in hydrolysis which is continued unless the desired free nitrogen concentration is obtained. It can take around 6 to 12 hours, which can be shortened when the temperature and pressure of the reaction is increased. The hydrolysate that is formed is then neutralized using NaOH (sodium hydroxide) and subsequently filtered and concentrated.

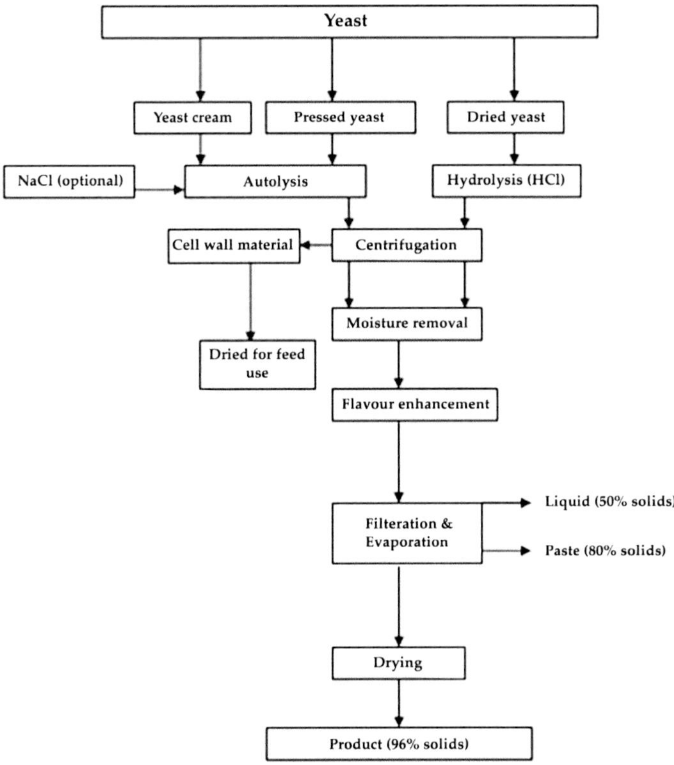

Figure 6.7: Outline of the process used for manufacture of yeast extract. (Source: *Murphy, R. A., & Horgan, K. A. (2005). Antibiotics, Enzymes, and Chemical Commodities from Fungi Fungi (pp. 113–143): John Wiley & Sons, Ltd.*).

6.7. YEAST THAT ARE ENRICHED IN SELENIUM

Because the demand for additives that are not genetically modified but organic has increased, it is necessary to produce 'organic vitamins and minerals' with methods that are economically feasible. To meet this demand, enrichment of the yeast species, *S. Cerevisiae* has been carried out by introducing minerals into it. One popular product is Sel-Plex which is a product of yeast that is rich in selenium. Microbes including yeast don't have the ability to differentiate between the two elements selenium and sulfur as they have similar chemical properties and are present in the same group. In fact, *S. Cerevisiae* is able to incorporate Selenium instead of sulfur into its metabolic products. When these cells are grown in a medium lacking sulfur and enriched with selenium (as selenate or selenite salts), they are able to utilize selenium instead of sulfur and synthesize selenium containing organic compounds. Most of these compounds are analogs of organic sulfur compounds including selenocysteine, selenomethionine, and selenocysteine. In addition, minute amounts of selenohomocystenine, selenocystathione, methylselenocysteine, *S*-adenosyl-selenomethionine, the selenotrisulfide, selenoglutathione, and various seleno-thiols (Figure 6.8). The pathways for biosynthesis of organo-sulfur compounds are also utilized for synthesis of organo-selenium compounds. The protein molecules in yeast cells can then synthesized by incorporation of these selenoamino acids.

In yeast cells, half of the dry weight has been shown to be composed of crude protein. This was characterized and shown to consist of 80% amino acids, 12% nucleic acids and 8% ammonia. Among the content of amino acids, *S. Cerevisiae* has 1.99% sulfur-containing amino acids (cysteine being 0.78% and methionine being 1.21%, representing 2600 and 2080 ppm, respectively, of the organically bound sulfur). It has been revealed through studies that while maintaining the biological activities of the proteins, 50% of their methionine content can be replaced by selenomethionine. This indicates that without compromising with protein synthesis and growth rate of yeast cells, selenium can replace up to 2000 ppm of yeast sulfur. Hence, if allowed to grow under appropriate conditions, yeast cells can grow in the presence of selenium when it is in excess of 2000 ppm. Figure 6.9 shows the process for the production of selenium yeast. There are several advantages in using selenium yeast instead of inorganic selenium. Selenium yeast poses less toxicity compared to inorganic selenium, therefore it is more environmentally-friendly. Moreover, there is lesser amount of selenium that remains unabsorbed and is excreted our through fecal route when yeast selenium is used. This prevents accumulation of toxic selenium into feces which is used as manure for farming. Increased bioavailability is economically beneficially. The feeds would be less expensive because it will lower the requirement for additional supplements in diets that are deficient in selenium.

6.8. CONCLUSION

The metabolism during the life cycle of fungi can be clearly divided into two stages; primary and secondary. The yeast metabolic processes have been commercially exploited using various techniques of biotechnology. It is highly valuable and economical for natural products to be obtained from fungi. these include production of fungicides, hormones, vitamins, enzymes, pharmaceutical drugs, antibiotics, plant growth regulators, and proteins.

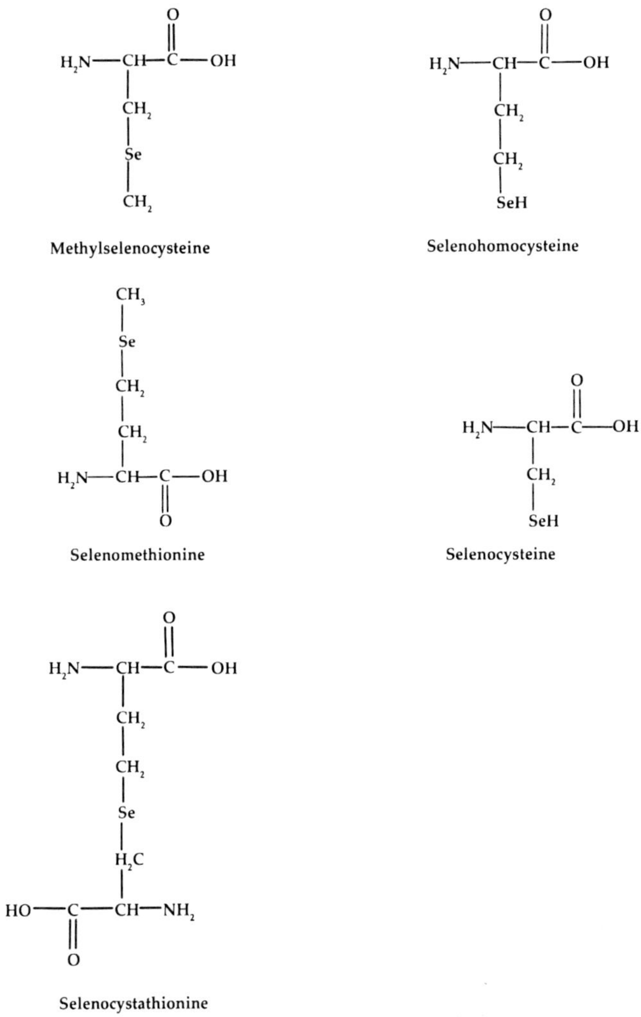

Figure 6.8: The variety of Selenium compounds present in Selenium yeast. (Source: *Murphy, R. A., & Horgan, K. A. (2005). Antibiotics, Enzymes, and Chemical Commodities from Fungi Fungi (pp. 113–143): John Wiley & Sons, Ltd.*).

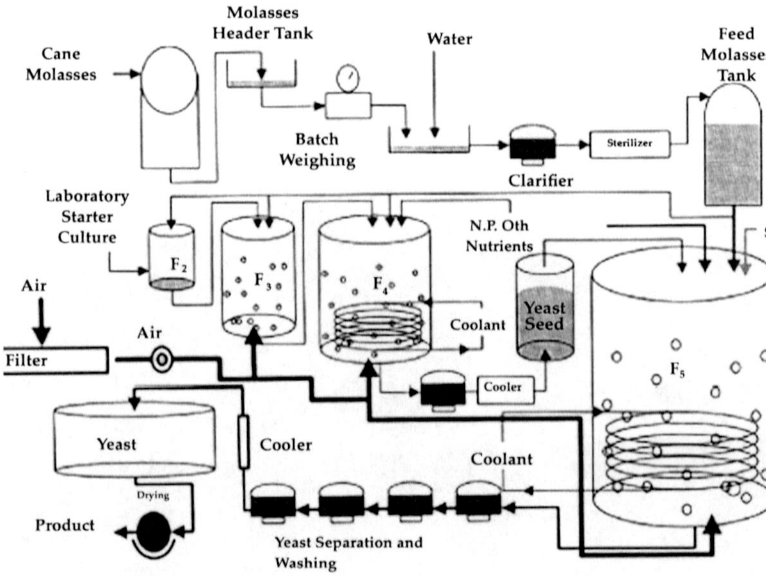

Figure 6.9: Industrial production of selenium yeast. (Source: *Murphy, R. A., & Horgan, K. A. (2005). Antibiotics, Enzymes, and Chemical Commodities from Fungi Fungi (pp. 113–143): John Wiley & Sons, Ltd.*).

7
CHAPTER

FUNGAL ENZYMES AND THEIR USES IN BIOTECHNOLOGY

CONTENTS

7.1. INTRODUCTION TO ENZYMES

Enzymes for the first time were detected as early as the nineteenth century. During precipitation of malt extract, starch was seen to be converted into fermentable sugars by a thermolabile substance. Because this process converted insoluble starch grains into soluble dextrins, the substance was termed diastase. Several other enzymes were identified by the mid-nineteenth century, including peroxidase, invertase, and polyphenol oxidase. The first commercial preparation of enzyme, termed Taka-diastase was produced by *Aspergillus oryzae* when it was grown on rice. There has been a dramatic progress since then in procedures for identification, extraction, and production of enzymes at commercial levels that are being applied in several industries.

Interestingly, half of the discovered and studied enzymes belong to fungi. The enzymes belonging to microbes offer several advantages over enzymes produced by plants and animals. These include higher catalytic activities, more yields, easy genetic manipulations and inexpensive production on large scale. In addition, they require the mild condition of pH and temperature and are biodegradable. Humans have long been using enzymes when utilizing biological systems for several purposes. For example, enzymes have been used for making cheese, producing antibiotics, brewing, and baking, manufacturing leather, linen, and indigo. They also have applications in the food industry (production of fruit juice, coffee, tea, and wine), textile industry, and paper industry, and detergents production.

7.2. INDUSTRIAL ENZYMES

The utilization of enzymes at industrial levels was not very successful during the early twentieth century. This was due to a lack of understanding about the activity of enzymes. It is only during the last three to five decades that the mechanism of working of enzymes and their kinetics have been elucidated. This improved knowledge has facilitated the industrial application of enzymes.

Enzymes that are significant in industrial applications are mostly hydrolytic depolymerases, such as lipases, carbohydrases, pectinases etc. Among these, proteases, and amylases that are employed in preparation of detergents are the most prominent ones. The remaining market is made up by utilization of carbohydrates, such as pectinases and amylases. To make high-fructose corn syrup, the enzyme glucose isomerase is used which are important at industrial level but do not come under the category of depolymerases.

Rapid evolution has taken place in the biotechnological industry for supply of enzymes market during the recent years. Currently, 4% and 5% annual growth rate and annual sales are observed in this sector which has crossed an amount of US$2 billion. The production and industrialization of enzymes has been revolutionized due to recent advances in protein and genetic engineering to variety of enzymes (Table 7.1). Industrial enzymes are categorized into three types; technical food and animal feed enzymes. The highest percentage of sales is for technical enzymes including amylase, protease, and cellulose. A vast number of industries utilize this category which includes starch, leather, textile, paper, pulp, detergent, and personal care industries. The second most marketed category of enzymes is those used in food industries. Enzymes like lipases and pectinases are used for production of wine, juices, dairy, fats, oils, and bakery products and for brewing. The third category, animal feed enzymes consists majorly of phytases, β-glucanases, and xylanases.

7.3. APPLICATIONS OF FUNGAL ENZYMES

There is a huge variety of fungal enzymes that are employed in the industry. In this section, we will be discussing the most prominent and economic examples of industries that utilize fungal enzymes.

7.3.1. Detergent's Industry

In 1913, the enzyme use originated when pancreatic enzymes were added into pre-soaking products. Soon after this, Brunus was introduced in the market which was the first enzymatic detergent. This was followed by several enzymes and formulations being utilized in various detergents for example mannanase which helps remove stains of foods that contain guar gum. The largest application of enzymes at industrial level is in detergents and the most prominent example of such enzymes is protease, which accounts for almost 30% of the market. Even though more expensive, washing powders used in the household often contain cellulases. These detergents remove fuzzy and small fibril from surfaces of fabric like cotton and enhance its color and appearance.

The increased popularity of enzymatic detergents is due to their cleaning capacity and because they do not contain phosphates and hence are more environmentally-friendly.

The following points need to be taken care of when selecting enzymes as detergent additives;

1. These enzymes should be stable and active within wide ranges of pH and temperature;
2. They should be sufficient and functional at levels as low as 0.4%;
3. They should be compatible for use with other components of detergents;
4. Their shelf-life should be long.

7.3.2. Biodiesel and Bioethanol

The need for production of biofuels like bioethanol and biodiesel has increased due to increased oil prices, increased environmental issues and fossil fuel resources being quickly used up. Currently, ethanol is the most popularly used biofuel and a lot of effort is being done to allow its production using cellulose and lignocellulose. Measures are being taken to reduce the cost of cellulose enzymes which are employed in processing of cellulose-containing materials.

Oils from rapeseed, soybean, vegetables, and animal fats are used for derivation of biodiesel which is an ester. Compared to petroleum, oil derived from vegetables is a lot cleaner as it doesn't release sulfur dioxide, thus reducing the soot production by almost one-third. Using lipase enzyme, trans-esterification reaction is carried out between oils and ethanol or methanol producing fatty esters. These products can then either be used as pure biodiesel or it can be mixed with the conventionally used diesel fuel.

7.3.3. Tanning Industry

In the leather industry, the skins and hides are de-haired using traditional methods involving mixtures of lime and sodium sulfide.

It is an unpleasant method which is a source of pollution. These animals skins are now being treated using methods that are more environmentally-friendly due to developments in biotechnology. Several mixtures of lipase and proteases are isolated from fungal species like *Aspergillus*. These mixtures of enzymes cause swollen roots of hair which results in easier removal of hair. The bating process in which the hair residues are removed by degradation of keratin and elastin is done. The leather produced as a result has higher quality compared to that manufactured by traditional methods.

7.3.4. Treatment of Effluent and Waste

Enzymes are now playing a key role in the waste treatment industry. Inactivated sludge and other aerobic waste processes, lipases are utilized in the breakdown of solids hence preventing blockages by fat. Waste feathers from poultry slaughterhouses are removed using proteases. Feathers constitute of 5% of the total poultry's body weight, which provides a high amount of proteins when used as feed after degrading their keratin structure. The feathers can be solubilized via hydrolysis by enzymes, addition of sodium hydroxide, followed by disintegration through mechanical methods.

A major source of environmental pollution is the large amount of wastewater that is released from textile industry. The effluent produced by textile is treated using biological procedures and with combination of chemical procedures. These systems are being continuously enhanced by improvements in biological techniques like bio-augmentation, immobilized microorganisms, and microorganism activity enhancements.

Table 7.1: INDUSTRIALLY used Fungal Enzymes, Their Source Species, and Their Applications

Enzyme	Organism	Application
α-Amylase	*Aspergillus niger, Aspergillus oryzae*	Starch processing and food industry
Chymosin	*A. niger*	Food industry
Cellulase	*Trichoderma viride, Trichoderma reesei*	Textile, pulp, and paper industry
Cellobiohydrolase	*T. viride, T. reesei*	Textile, pulp, and paper industry
Glucoamylase	*Aspergillus phoenicis, Rhizopus delemar*	Starch processing industry
Glucose oxidase	*A. niger, A. oryzae*	Textile industry, biosensor
Laccase	*Trametes Versicolor*	Textile, pulp, and paper industry
Lipases	*A. niger, A. oryzae*	Food and detergent industry
Pectin lyase	*T. reesei*	Food industry
Proteases	*A. niger, A. oryzae, R. delemar*	Food and detergent industry
Phytase	*A. niger, A. oryzae*	Food industry
Rennin	*Mucor miehei*	Food industry
Xylanases	*T. reesei, Trichoderma konignii, A. niger*	Textile, pulp, paper, and bakery industry
Amyloglucosidase	*A. niger*	Starch syrups, dextrose, foods
Invertase	Yeast species	Confectionary industry

(*Source: Meyer, V. (2008) Genetic engineering of filamentous fungi–progress, obstacles, and future trends. Biotechnology Advances, 26, 177–185*)

7.3.5. Processing of Food

Enzymes are utilized most significantly in the food industry. These enzymes vary in the range from lipases and catalases used in making and ripening of cheese to pectinases for brewing wine and processing of juices, α-amylases for liquefaction of dextrans and proteases for tenderizing meat. The advantages of employing enzymes for food processing are as follows:

1. These catalysts are very specific and controllable and formation of undesirable side products is avoidable;
2. In comparison with methods involving chemicals, enzymes require milder conditions of pH, temperature, and pressure;
3. There is no problem of toxicity when enzymes are used as enzymes are biodegradable;
4. It is easy to immobilize them on solid substrate which makes easier processing and purification of products. This also lowers the chances of allergenicity.

Lipases are used to produce lean fish, refine flavor of rice, and to modify milk from soybean. Additionally, they are used for enhancing flavor of cheeses, and producing products of cheese including soft cheeses. Proteases are used in tenderizing meat, hydrolyzing proteins in connective tissue and fibrous proteins in muscles. Enzymes are either injected into the blood or directly into the meat. To improve dough for bread, currently, proteases, amylases, and cellulases are employed. As a results, the freshness, flavor, volume, and texture of bread is enhanced. Moreover, it the machinability of the bread dough is also improved. Current studies are being carried out to study staling of bread and the role of enzymes such as xylanase and α -amylases to prevent this process.

7.3.6. Clarification of Fruit Juices

Juices from and vegetables are extracted and clarified using macerating enzymes which includes cellulases, hemicellulases, and pectinases. Fruits that contain starch, such as those from apples are prevented from forming haze using enzymes like laccases, amyloglucosidases, and α -amylases. Pulp of fruits like apples, bananas, and grapes, when treated with pectinases, increased the volume of juice obtained. As more fruits and vegetables are becoming applicable, there is an increasing demand for such enzymes.

There has been an ease in peeling citrus fruits by developing the process called vacuum infusion. This method was developed when it was observed that preparations of impure naringinase into peel of grapefruit which was actually employed to lower its bitterness also resulted in softening of albedo, the white part of the peel of citrus fruits. This process is also employed in production of pickles where softening occurs while products are fermented and stored. In food technology, infusion of enzymes to influence sensory attributes of vegetables, fruits, and other foods have a huge potential.

7.3.7. Feeds for Animals

A major part of the agro-industrial activities in the world is that of animal feed industry which produces feeds of more than 624 million tonnes annually having worth of more than US$50,000,000. This industry has gone through many alterations during the recent years since the industry and its consumers have studied the production of compound animal feeds, rearing of animals and the environmental effects of systems used for animal husbandry. Production has been enhanced greatly through employment of enzymes like cellulose, amylase, β-glucanase, and protease into feeds. One important example is of diets based on barley which are used for poultry industry which are produced by the help of xylanase and β-glucanase enzymes. This results in feeds production that are easier to digest and absorb and which help increase weight of broiler chicken and egg-laying hens.

Another positive effect of using enzymes is digestibility of non-starch polysaccharides (NSPs) by piglets that are fed with barley. When enzymes are incorporated into animal feeds, they exhibit several positive effects on performance of animals. The names of enzymes and their substrates along with their positive effects are summarized in Table 7.2.

When diets are supplemented with selective enzyme activities, there is a reduction of pollutants present in animal wastes. Especially in case of phosphorus in the diet which remains undigested in monogastric animals. Addition of phytases that are produced by microbes can help reduce animal waste and enhance utilization of nutrients. Several novel fungal enzymes have been recently identified and have specific activities greater than 4 to 50 times compared to previously used enzymes.

Table 7.2: Beneficial Effects of Fungal Enzymes and Their Substrates

Enzymes	Substrate	Beneficial effects
β-Glucanases	Mixed-linked glucans	Reduction in viscosity of digesta
		Reduction in incidence of pasted vents
Xylanases	Pentosans	Improvements in litter characteristics
Cellulases	Cellulose	Reduction in dirty egg problems
		Reduction in viscosity of digesta
Phytases	Phytic acid	
		Promotes a more comprehensive digestion of vegetative matter
Proteases	Proteins	Removal of anti-nutritional effects of phytic acid
Amylases	Starch	
		Supplementation of endogenous proteolytic and amylolytic capacity to benefit young or sick animals whose digestive function may not be operating maximally.

(Source: Walsh, G. (1998) Biopharmaceuticals: Biochemistry and Biotechnology, John Wiley and Sons, Inc., New York.)

The stored form of organic phosphate in plants is phytic acid which comprises around 60% of the total phosphorus in cereal grains. This phytic acid is not hydrolyzed properly by broilers. This compound binds to and thus prevents digestion and absorption of other important components of diet including zinc, magnesium, calcium, and proteins. This results in excretion of phytic acid in the form of manure posing problems to the environments around livestock production areas. When phytase is used, it degrades phytate which lowers the risks to the environment. In addition, this category of enzymes provides other advantages; it destroys the anti-nutritional characteristics of phytic acid, the requirement for supplements containing inorganic phosphate is reduced, feces contain lesser amount of phosphates which leads to reduced levels of phosphorus added into the environment. Phytase enzyme is being used specifically in areas with environmental issues but it can be extended to more areas if its utilization can be made more efficient.

7.4. FUTURE ASPECTS OF ENZYMES USED INDUSTRIALLY

It is expected that industrial enzymes, in the near future, will be used more prevalently and their utilization will be expanded to areas like food processing, chemical conversions, biopulping, additives of animal and food feeds, carbohydrate conversions, cleaning, and detoxification of toxins in the environment. Initially, the production of enzymes on a large scales was limited to only 20 enzymes but this number increased dramatically as there were developments in molecular genetic tools. Particularly, the genes coding for non-abundant enzymes are cloned into heterologous hosts which overexpress the proteins. This method helps in the expansion of industrial utilization of enzymes.

7.5. EXAMPLES OF FUNGAL ENZYMES

Several fungal genera including *Penicillium*, *Rhizopus*, and *Aspergillus* are used in commercial production of enzymes like cellulose, protease, lipase, amylase, phytase, and xylanase. In this section, we will discuss these enzymes in detail.

7.5.1. Protease

This class of proteins is broadly classified as enzymes that hydrolyze protein molecules by breaking the peptide bond (CO-NH). These protease enzymes differ in their optimal pH. Generally, bacterial proteases function in alkaline conditions while fungal proteases function in acidic conditions. All living organisms consist of proteolytic enzymes and account for almost 2% of the total proteins present in them. Various industries employ proteases and the huge variety of the available proteases has enabled their exploitation in both biotechnological and physiological applications.

Proteases are produced naturally by animals, plants, and microorganisms. But those produced by microorganisms like bacteria are preferred because their biochemical diversity is broader and they are easier to be manipulated genetically. Therefore, microbial proteases are well-characterized and their molecular properties have been well-understood.

Commercial production of proteases is done through utilization of the filamentous fungi like *Aspergillus*. In comparison to bacterial proteases, the fungal proteases belong to a wider variety and are active in broader ranges of pH.

7.5.1.1. Types of Proteases

Depending on their action site, proteases are divided into two categories; endopeptidases and exopeptidases. The latter, also called as peptidases, are further grouped into two types depending on whether they excise an amino acid from the N-terminus or the C-terminus of proteins. They specifically act on dipeptide substrates. Some exo-acting peptidases excise dipeptide units from either the C-terminus or the N-terminus of the proteins. Endopeptidases, also called as proteinases, on the other hand, break the peptide bonds that are present internally within a peptide chain. This class of enzymes is further divided into four groups depending on the functional amino acid present at their active sites. These include cysteine, serine, aspartate, and metalloproteases (Table 7.3).

Enzymes that catalyze the cleavage of amino acid units from protein or peptide's N-terminus are exo-acting peptidases which are called aminopeptidases. This category of protease is produced widely by animals, plants, and several microorganisms (such as fungi and bacteria).

The proteases that cleave single amino acids or dipeptides from C-terminus of proteins or peptides are called as carboxypeptidases. Similar to aminopeptidases, carboxypeptidase are further divided on the basis of amino acids present at the enzyme's active site; cysteine, serine, and metallocarboxypeptidases.

As the name implies, serine proteases have serine residue at their active site. They are strong endopeptidases with low specificities. They function by transferring an acyl group from substrate to a functional group present on the enzymes. These enzymes are very important and are prevalently produced by eukaryotes, bacteria, and viruses. Serine proteases also come under the category of exopeptidases, oligopeptidases, and omega peptidases. These enzymes are divided further into 20 to 30 families depending on their structural similarities.

In filamentous fungi, aspartic proteinases are produced abundantly and they have low optimum pH. As the name suggests, they have aspartate residue in their active site. This group of enzymes is resistant to serine protease inhibitors, chelating, and thiol-group reagents. They are similar to digestive enzymes present in animal, such as pepsin and rennin. Therefore, they can replace animal proteases and in manufacture of cheese.

Table 7.3: Proteases classes. Solid Circles Represent Terminal Amino Acids, Open Circles Represent Internal Amino Acids, and Stars Represent Blocked Termini of Polypeptides. Arrows Signify the Cleavage Sites of the Enzymes

Protease	Mode of Action	EC no.
Exopeptidases		3.4.11
Aminopeptidases	•↓-o-o-o-o—	3.4.14
Dipeptidyl peptidases	•-•↓-o-o-o—	3.4.14
Tripeptidyl peptidase	•-•-•↓-o-o—	3.4.16–3.4.18
Carboxypeptidase	—o-o-o-o-o↓-•	3.4.16
Serine-type protease		3.4.17
Metalloprotease		3.4.18
Cysteine-type protease	—o-o-o-o↓-•-•	3.4.15
Peptidyl dipeptidases	•↓-•	3.4.13
Dipeptidases	*-•↓-o-o—	3.4.19
Omega peptidases	—o-o-o↓-•-*	3.4.19
Endopeptidases	—o-o-o↓-o-o-o—	3.4.21–3.4.34
Serine protease		3.4.21
Cysteine protease		3.4.22
Aspartic protease		3.4.23
Metalloprotease		3.4.24
Endopeptidases of unknown catalytic mechanism		3.4.99

(Source: Rao, M.B., Tanksale, A.M., Ghatge, M.S. and Desphande, V.V. (1998) Molecular and biotechnological aspects of microbial proteases. *Microbiology and Molecular Biology Reviews*, 62, 597–635).

Cysteine proteases, divided into 20 families are majorly produced by eukaryotes and prokaryotes but they are less abundantly produced by fungi. These proteases function depending of a catalytic dyad made up of cysteine and histidine residues which are arranged differently in different families. These enzymes function through a mechanism very similar to serine proteases.

The most diverse class of proteases is metalloproteases which consist of up to 25 families. These enzymes generally have a metal atom within their structure and they are optimally active at a pH around 7. Because their stability is dependent on the presence of calcium ions, several chelating agents can be used for their destabilization and inactivation.

7.5.1.2. Proteases Applications

Proteases have been used in industry of food since ages. For example, baking, production of alcohol and tenderization of meat were all done with the help of enzymes produced from microorganism by ancient Greeks. Proteases have been employed in several industries including pharmaceutical, detergents, diagnostics, chemicals, and animal feeds. Among these, the detergent and feed industries use proteases most extensively.

The popularity of proteolytic enzymes in detergent industry is because they offer several advantages. They are stable over a broad range of temperatures and exhibit optimum activities in alkaline conditions. Detergents used in laundry that have protease additives help remove stains that contain proteins like blood, milk, gravy, and keratin. Due to the recent crisis of energy and the urge to conserve of energy, there is a continuous search for novel proteases. Interestingly, there have been efforts made for production of proteases using substrates like crab shells, shrimps, soybean, and fish waste.

In the dairy industry, proteases are employed in cheese manufacture because they are able to clot milk. Traditionally, cheese was made using rennet isolated from calves' stomach. Due to increasing demand, there has been a shortage of rennet which urged the researchers to look for alternate sources. A replacement for calf rennet is recombinant chymosin which is produced by strains of fungi such as *Mucorpusillus* and *Mucor miehei*. Additionally, cheese ripening is done through utilization of protease isolated from *Penicillium caseicolum* and *Penicillium roqueforti.* In the baking industry, the texture of bread and uniformity of dough is achieved with the help of proteinases produced by *A. oryzae*.

A wide variety of proteases with different specificities are used as effective therapeutic agents. The therapeutic applications of collagenases in increasing for dosage forms that are slow-releasing. The deficiency of lytic enzymes is corrected by using proteolytic enzymes produced by *A. oryzae* which help in digestion. The diverse range of proteases found in microbes allows their exploitation in various ways.

7.5.2. Cellulase

The major component of cell wall is cellulose. It is most prevalently found polymer that is renewable. It is made up of residues of glucose that are linked by 1,4-β-glycosidic bonds. Hydrogen bonding helps link cellulose molecules resulting in formation of large lattice-like structures. Because it is composed of a complex structure, degradation of cellulose is done using multiple enzymes. As a result, microorganisms that secrete cellulolytic molecules help degrade cellulose.

Studies have shown that cellulose breakdown requires the collective action of β-glucosidases, exoglucanases, and exoglucanases. Cellulose, a type of endoglucanase, cleaves cellulose at random positions to produce oligosaccharides. Glucose and cellobiose exoglucanases cleave β-1,4-D-glucosidic linkage within cellulose molecules that release cellobiose from the non-reducing end. β-glucosidases, on the other hand, degrade cellobiose into glucose (table 7.4). Cellulose has a huge potential as a renewable source of energy because it is produced as a major side-product through activities in nature and man-made.

Table 7.4: Types of Cellulose Present in Fungi and Their Mechanism of Action on Cellulose Molecules

Type of enzyme	EC code	Synonym	Mode of Action
endo-(1,4)-b-D-Glucanase	EC 3.2.1.4	Endoglucanase or endocellulase	–G—G—G—G— ↑ ↑ Cleaves linkages randomly
exo-(1,4)-b-D-Glucanase	EC 3.2.1.91	Cellobiohydrolase or exocellulase	G—G– G—G—G— ↑ Releases cellobiose either from reducing or non-reducing end

exo-(1,4)-b-D-Glucanase	EC 3.2.1.74	Exoglucanase or glucohydrolase	G—G—G—G—G— ↑ Releases glucose from non-reducing end
b-Glucosidase	EC 3.2.1.21	Cellobiase	G—G, G—G—G—G ↑ ↑ Releases glucose from cellobiose and short-chain xylooligosaccharide

(Source: *Bhat, M.K. and Bhat, S. (1997) Cellulose degrading enzymes and their potential industrial applications. Biotechnology Advances, 15, 583–620.*)

Cellulases are synthesized by a several bacteria and fungi that are both anaerobic and aerobic, thermophilic, and mesophilic in nature. The enzymes that are isolated from aerobic fungal strains are well-studied. These include *Penicillium pinophilium*, *Fusarium solani*, *Sporotrichum pulverulentum*, *Trichoderma reesei* and *Trichoderma viride*.

Because of their ability to breakdown lignocellulose into glucose and other soluble sugars, research on cellulolytic enzymes initiated as early as 1950s.

Hemicellulase and cellulases were first applied in the field of biotechnology during early 1980s, and subsequently, they were used in the food, textiles, pulp, paper, and laundry industries.

Celluloses are known to a potential source of energy that is renewable and most prevalent. During World War 2 there was an alarming situation in South Pacific due to diminishing of cellulosic material such as tents, clothing, and sandbags. The US army set up organizations to look for solutions to this problem immediately. This research resulted in identification of *T. viride* and *T. reesei* strains that were shown to be exhibit hyper-cellulolytic activities. *T. reesei* strains were selected, isolated and characterized. As a result of this research, T. reesei was utilized to produce cellulose in large amounts.

Because fungi have a key role in biotechnology, several enzyme systems have been recognized that exhibit cellulolytic activities. These enzymes are used in food industry for producing juices where cellulases and other enzymes help to extract and clarify juices. Cellulase enzymes are also used in farming industries to provide supplementation to livestock which helps improves their performance as well as yield. Moreover, cellulases have also been employed in the textile industry as they are able to modify cellulose-containing fibers in a desirable and controllable way. In the textile industry, utilization of cellulases began only approximately a decade ago but within this time they have gained so much popularity that they have become the third largest enzymes groups used. In the paper and pulp industry, cellulose and hemicellulose enzymes are used in combination to allow biobleaching and biomechanical pulping.

Cellulolytic enzymes have a major application in formation of fuel and ethanol using biomass. Contrary to fossil fuels, ethanol is renewable source of energy which is formed by sugar fermentation process. it is commonly used as a fuel instead of gas in the USA. Cornstarch is used for an annual production of 1 billion gallons of ethanol using procedures employing yeast. Using this process on grains with high value yields some products used as animal feeds and other by-products. Corn has been used for production of ethanol which is used as oxygenated fuels or gasohol since the 1980s. Compared to fossil fuels, the production cost of ethanol is still much greater. This industry is already mature; therefore, chances of further improvements are limited. However, ethanol can be produced at low costs if lignocellulose-containing materials are used as raw materials. These include wood, animal manure, grasses, sawdust, and crop residues such as stems and stalks. If such technologies can be developed and implemented, it would need for importing oil, enhance quality of air and help dispose solid animal waste in a natural way. The efficiency can be enhanced and cost of production of ethanol reduced through development in the organisms and processes utilized for bioconversion of lignocellulose to ethanol relative to petroleum.

A major advantage of using cellulase enzymes for ethanol production is its low cost. To further enhance the applicability of cellulases, recombinant technology can be used which will allow production of enzymes by microorganisms that degrade cellulose and prevent build-up of inhibitory products hence reducing the need for cellulases from fungal strains.

7.5.3. Xylanase

Xylan is an organic compound which is prevalently pound in plant cells and can be degraded to release important end products. Its structure consists of linear chains of xylopyranose residues which are linked by β-1,4 bonds and their main chain is bound to substituents through ester or glycosidic bonds. Because xylans are heterogeneous, their degradation is carried out by multiple enzymes including esterase, β-xylosidase, ndoxylanase, α-arabinofuranosidase, β-glucuronidase. The breakdown of xylan is initiated by endo-1,4-β-xylanase which produces xylooligosaccharides and xylose. In angiosperms and gymnosperms, xylan is an abundant component of their hardwood and softwood, respectively. It is also found in annual plants at quantities of <30%. Generally, xylan is present in plants' secondary cell walls.

Several organisms are able to produce xylanases, such as, fungi, bacteria, algae, anthropods, gastropods, and protozoa. *Trichoderma* and *Aspergillus* are the most prominent fungal strains that are xylanase producers. Thee microorganisms live in a wide range of habitats including places where plants grow and deteriorate, and ruminants' stomachs.

Even though xylanases are produced by various organisms, they are preferably commercially produced by filamentous fungi because they produce high yields of enzymes with high specificity and they propagate in mild environmental conditions. Xylanases isolated from fungal strains are able to function at temperatures around 40 to 60°C and pH in the acidic range. Some xylanases, however, are able to tolerate in extreme conditions. Such enzymes majorly belong to families 10 and 11. From the Antarctic environment, some psychrophilic fungi have been isolated which include *Alternaria alternate, Phome* sp., and *Penicillium.*

sp. Xylanases produced by these strains all have low optimum temperatures and high catalytic activities and poor stabilities.

Developments in stabilizing the enzymes under alkaline and acidic conditions, enhancing their activities and thermostabilities have enabled to expand the repertoire of commercially employed fungal xylanases. During the last decade, scientists have had keen interest in utilization of xylan and xylanases in biotechnology. Among the various products of xylan, furfural obtained from agricultural residues and xylitol derived from wooden residues have high significance.

The xylanolytic activities have industrial significance and have utilization in food and beverages industries. They are, for example, used in clarification of wines and juices, and for enhancement of dough properties and bakery products' qualities. Moreover, they are used in poultry industry for enhancing efficiency of feeds and for improving nutritional values of green feed and silage. It is sometimes also used in brewing industries for improving filterability of worts in order to reduce haze in the end product, in extraction of coffee for production of soluble coffee.

Since there have been restrictions in using chlorine to bleach paper, xylanases have gained popularity in the paper and pulp industry. Using chlorine releases harmful bi-products that exhibit toxicity, mutagenicity, bioaccumulation which results in several dangerous alterations in the biological systems. To bleach pulp, the traditional methods have been replaced by biobleaching which involves the use of microorganisms and enzymes. Xylanases isolated from microbes breakdown the backbone of xylan structure hence rendering them more susceptible to bleaching chemicals like chlorine dioxide, thus, reducing the amount of bleaching chemicals that are needed to allow production of pulps with the desired brightness values. Moreover, xylanases are used for fibrillation of pulp, allowing low beating times lowering the consumption of energy. It also help increase the freeness in recycled fibers.

Production of xylanases has increased significantly recently. This is evident from 468 patents that have been published in United State Patent and Trademark Office since 2001. The future developments in xylanase utilization have been urged due to increasing environmental issues regarding greenhouse gases.

7.5.4. Amylase

Currently, amylases are most prominent in biotechnology and have several applications in industries such as food, textiles, paper, and fermentation. It have become so prevalent that amylases produced by microbes are used as a replacement of chemical hydrolysis of starch in industries that process starch. Amylases are of three types; α -amylase, β-amylase, and glucoamylase. Starch is a polymer made up of two types of polymeric units; amylopectin and amylose. These units are made up of α -D-glucose residues. In amylose, the glucose units are joined together by α -1,4-glycosidic bonds resulting in a linear molecule which are arranged in a double helical structures that has 6 molecules of glucose in each turn. On the other hand, amylopectin is a branched molecule which has 20 to 25 glucose molecules arranged linearly and α -1,6 glycosidic bonds (4–6%) at branches.

Amylopectin is one of the largest molecules found in nature and can have molecular weights greater than the ranges of 10^8. Grains of starch are present in cells in irregular or rounded shapes. When they are unprocessed, they have lengths between 1 and 100 μm with their structure being stabilized by hydrogen bonds.

Amylases are produced prevalently by several fungal species. Even though several microbial organisms synthesize amylase, selecting strains that are able to produce high yields of amylase that are acceptable at industrial level is challenging. Amylolytic enzymes are commercially produced using two main processes; solid-state fermentation and submerged fermentation. Recently, solid-state fermentation has become more prevalent for synthesis of enzymes used for saccharifying starch.

Industrially used amylases are thermostable. As a result, microorganisms that are thermophilic are especially important for obtaining thermostable amylases.

7.5.4.1. α-Amylase

This category of enzymes, also called as glycogenase, *endo*-1,4-α -D-glucan glycohydrolase, or 1,4-α -D-glucan glucanohydrolase, is produced extracellularly. It cleaves the 1,4-α -D-glycosidic bonds within the linear amylase chains at random positions. These enzymes are endo-acting and are able breakdown chains of amylose internally and are categorized on basis of their mechanism of action and the end products achieved. Some enzymes can produce free sugars and are called 'saccharogenic' while some just liquefy starch and are unable to yield free sugars and are termed as 'starch liquefying.' α-amylase are able to degrade long chains of starch carbohydrates to produce maltose and maltotriose from amylose units and glucose and limit dextrin from amylopectin units. Because of its ability to hydrolyze starch at random places, these enzymes are more effective compared to β-amylase.

Commercially, α-amylases are synthesized from both bacterial and fungal strains such as *Bacillus* and *Aspergillus*. The enzymes produced by bacteria are thermostable and are therefore more prevalently utilized compared to fungal amylase. The bacterial strains *Bacillus amyloliquefaciens* and *Bacillus licheniformis* are generally used for commercial use. Using thermostable α-amylase offers several advantages such as minimizing contaminants, reduction of reaction times, and reduction in polymerization of D-glucose to iso-maltose. α-amylases have also been produced by filamentous fungi. Thermostable α-amylases have been shown to be abundantly produced by thermophilic fungal strain *Thermomyces lanuginosis.* A-amylase that is tolerant to acidic conditions is produced through solid-state fermentation by *A kawachii.* When producing α-amylase using *Pycnoporus sanguineus,* solid-state fermentation proved to be more favorable compared for submerged fermentation. A *Rhizopus* strain has also been shown to produce abundant amounts of α-amylase that is thermostable through solid-state fermentation process.

7.5.4.2. *β-Amylase*

This type of amylase, also called as glycogenase, saccharogen amylase or 1,4-β-D-glucan maltohydrolase, is generally produced by plants but some strains of microbes are also reported to be β-amylase producers. This type of enzyme is exo-acting and it breaks down ends of non-reducing chains of amylopectin, amylose, and glycogen. It does so by breaking glycosidic bonds which results in formation of maltose. This enzyme is unable to hydrolyze α-1,6- glycosidic bond present in amylopectin, therefore, incomplete breakdown of starch results in products that are around 50% maltose and a β-limit dextrin. In contrast to other types of amylases, β-amylases produced by microbes are rarely used at commercial levels. Bacterial β-amylases have been well-characterized even though some fungal strains like *Rhizopus* are also shown to produce this enzyme.

7.5.4.3. *Glucoamylase*

Glucoamylase are also known as amyloglucosidases, glucan 1,4-α-glucosidase, *exo*- 1,4-α-glucosidase, lysosomal α-glucosidase or 1,4-α-D-glucan glucohydrolase. They are able to cleave glucose units from amylose and amylopectin at their non-reducing ends. Contrary to α-amylases, these enzymes are able to break 1,6-α -glucosidic bonds which are located at branch points of amylopectin, even though it is able to do so at a slow rate. Glucoamylase hydrolyzes starch to produce glucose, maltose, and β-limit dextrins. Most of the glucoamylase have multiple domains; they contain catalytic domains which are bound to domains that bind to starch. This category of amylase is found prevalently among several microbial strains. Interestingly, glucoamylases are abundantly produced by filamentous fungi and are therefore used for production of fructose and glucose syrups. For commercial use, these enzymes are produced from *A. niger* which is cultured in solid-state on waste products of agro-industrial processes such as wheat, bran, rice bran, rice husk, gram, wheat, and corn flour, tea waste and copra waste etc. Sometimes, isolated of *A. awamori* and *Rhizopus* are also employed for commercial production of glucoamylase. On the basis of their soluble-starch digestive glucoamylase (SSGA) and raw-starch-digestive glucoamylase (RSGA) activities, strains of *Rhizopus* are categorized into four types. Yeast cultures that are able to produce glucoamylases include *Saccharomyces cerevisiae*, *Saccharomycopsis fibuligera*, *Candida fennica*, *Candida famata* and *Endomycopsis fibuligera*.

7.5.4.4. Amylase Applications

Agricultural products that are rich in starch are useful substrates for microbial conversion into fuels, feeds, and chemicals. Among these, the most important substrates are rice, wheat, corn, potato, oats, and cassava which are made up of 60 to 75% starch (wt/wt).

Insoluble grains of starch are first dispersed into an aqueous solution and subsequently hydrolyzed partially by employing amylases that are thermostable in order to achieve starch liquefaction. At commercial level suspension obtained through starch, liquefaction contains 35% (w/v) starch which is gelatinized to produce a high viscosity suspension. To reduce viscosity, thermostable α-amylases are used for hydrolyzing the starch partially. Prior to employment of thermostable amylases, liquefaction of starch was done through hydrolysis by hydrochloric or oxalic acids for 5 mins at 140 to 150°C temperatures. These harsh environmental conditions are not required since the use of thermostable amylases which work in environmentally more favorable and mild conditions. Moreover, this process has resulted in a reduction in release of side-products and recovery and refinement of products has become more cost-effective.

To produce ethanol, starch has proved to be a high-yielding resource. Several raw materials have shown to be useful for its production. However, production of ethanol through direct fermentation of raw materials by yeast is generally not possible. In such cases, it is necessary to use saccharifying enzymes from fungi to enable fermentation. Starch is first milled and cooked to allow its hydration and gelatinization. Subsequently, it is then pretreated with amylolytic enzymes which results in its hydrolyzation into simpler sugars. These simple sugars are fermentable by yeast and are thus converted into ethanol which can then be used in many different ways.

The most significant industrial use of starch conversion is for the production of high-fructose corn syrups or sweeteners. In foods and beverages, conventionally used sucrose syrups can be replaced by this product as it is very sweet. For production of high-fructose corn syrups, corn is first milled to form cornstarch. This is followed by treatment with α-amylase to yield small oligosaccharide chains. These short chains of sugars are further broken down to form glucose through glucoamylase produced by *Aspergillus* strains. Finally, the product is converted into 42% fructose by xylose isomerase enzymes the resultant fructose-rich syrup is further purified to yield high-fructose corn syrup that consists of around 90% fructose.

Other than ethanol production, amylases are also employed in making bread. Wheat flour, which mainly composed of starch, when exposed to amylases produces dextrins with low molecular weights. These dextrins are utilized by yeast to raise the bread and impart flavors into it. Addition of amylases into bread dough helps make the commercial bread-making process faster and more practical.

7.5.5. Lipase

Lipases, also known as triacylglycerol acylhydrolases, are hydrolytic enzymes which are prevalently found in animals, plants, and microorganisms. Typically, they are utilized for breaking down triglycerides into various products; diglycerides, monoglycerides, glycerol, and fatty acids. In addition, lipases are also used for catalyzing esterification, transesterification, and interesterification in a medium that is inaqueous. Commercially, there are around 35 different lipases available. One major advantage of using lipases is that they can be employed in mixtures containing water, organic solvents, or even both. As a result, these enzymes are used in a versatile manner in a wide range of industries including cosmetic, detergent, textile, paper, pharmaceutical, and leather.

Many fungal strains produce lipases that are well-characterized and exhibit various properties on the basis of their source. These properties include their thermostability, positional specificity, activity optima and fatty acid specificity. Fungal lipases that are used commercially are obtained from *Rhizopus* sp., *Rhizomucor miehei*, *Geotrichum candidum*, *Pichia burtonii* and *Candida cylidracae*. The most prevalently used lipase is extracted from *Rhizomucor miehei*. Moreover, this lipase also serves as the model for representing structure of various other lipases.

Lipases from microbes are biotechnologically significant enzymes. It is expected that their market is going to grow in future, especially in the detergent and cosmetics industry. Currently, lipases form the largest commercially used group of enzymes which form a market of around billion dollars and have a vast application. Basically, lipases are employed in two major ways; they are used for manufacturing fine chemicals or they are used as catalysts for manufacturing ingredients of food. Recently, new application of lipases have been discovered including production of biodiesel, biopolymers, and enantiopure pharmaceuticals, flavors, and agrochemicals. The industrial production of several products which is traditionally done by chemical processing of fats and oils can now be generated at much quicker rates with greater specificity under milder conditions using lipases.

Their key properties which make them very successful as biocatalysts are their broad specificity for substrates, their high selectivity for region type and stereotypes, their high stability in organic solvents, and their ability to function in mild environments. Their activity under mild conditions help reduce the effects of reaction conditions on products are reactants that are sensitive.

In dairy industry, lipases are used for hydrolyzing milk fat. In addition, they are also used for enhancing flavors of cheese, improving the rate of cheese ripening, and for lipolysis of cream and butterfat. Hydrolysis of milk fat by lipases is applicable on many dairy products including soft cheese which has specific characteristic flavor. For instance, some lipase action results in formation of fatty acids that contain short chains (4 or 6 Carbons) which produced a tangy, sharp flavor, while fatty acids that contain medium chains (12 or 14 carbon) induce a soapy taste to the final product.

Lipolytic enzymes can promote generation of free fatty acid chains and this ability is employed in chemical reactions for various purposes. For example, their action can produce acetoacetate, β-keto acids, methyl ketones, esters, and lactones which help enhance flavoring of the products. Preparations of fungal lipase enzymes, including those from *M. miehei*, *A. niger* and *A. oryzae*, have been manufactured by the cheese industry.

Due to their structure, some fats are very valuable in the food industry. Trans-esterification by lipases can help use cheaper oils in the products. For instance, the mid-fraction of palm oil is used to produce cocoa butter. Moreover, trans-esterification reactions carried out in the presence of lipases in organic solvents are now becoming increasingly popular at industrial level. They are employed for synthesis of a substitute for human milk fat, polyunsaturated fatty acids that have pharmaceutical importance, substance equivalent to cocoa butter and biodiesel from vegetable oils.

Lipases belonging to *M. miehei* and *Candida Antarctica* are used to esterify free fatty acids without organic solvents and also for trans-esterify isopropylidene glycerols with fatty acid methyl esters in hexane. To produce vegetable oils rich in omega-3 polyunsaturated fatty acids such as those from corn, sunflower, peanut, olive, and soybeans, lipases from *M. miehei* is immobilized to allow inter-esterification reactions.

Tallow is also hydrolyzed at industrial level by using lipases. It offers several advantages including; reduction for requirement of heat, thus reducing the consumption of fossil fuels and as a result, lowering the negative impact on environment. Moreover, due to low temperatures employed, there is lesser break down of unsaturated fatty acids and as a result, highly unsaturated oils can yield pure, natural fatty acids. The un-degraded polyunsaturated fatty acids may be significant for food preservation for production of additives for foods like mono- and diglycerides. In the textile industry, lipases have many roles. For example, they are commercially prepared and used with α-amylases to allow removal of size lubricants and for desizing fabric made up of cotton. The most prominent commercial utilization of lipases is in production of detergents used industrially as well as in the household. Typically, detergents that are in use today contain a mixture of several enzymes including amylase, lipase, cellulase, and protease. Using these detergents lowers the impact on the environment as they require low temperatures for washing, they are sufficient in small amounts, eliminate the need for other chemicals in detergents, they are biodegradable, do not release any toxic residues, they cause no problem to sewage treatment systems and they do not cause any harm to aquatic life.

Liopolase, the first commercially produced lipase is produced from the fungus *Thermomyces lanuginosus.* Because the natural enzymatic yield produced by the strain was low, its gene was cloned and expressed in another fungal strain, *A. oryzae.* This preparation of enzyme is very commonly used in industries like textile and detergents. Another example for lipase preparation which is significant at commercial level is the recombinant *Humicola* lipase which is produced through heterologous expression in *A. oryzae.* This preparation is important in detergents industry.

7.5.6. Phytase

Because this enzyme is not present in digestive tracts of monogastric animals, it has found prevalent use as an additive in animal feeds used in the animal industry including those of pigs and poultry. Phytase is able to carry out hydrolysis of non-nutritional factor, phytate (or phytic acid) into inositol and inorganic phosphorus.

Phosphorus is one of the essential elements necessary for all living organisms. It plays vital roles in growth, development, structure of skeleton, and various metabolic pathways. Diets that are deficient in phosphorus has many negative effects of performance of livestock, such as, reduction of appetite, malformation of bones, and reduced fertility. To overcome this problem, it is necessary to supply the animals with an external source of phosphorus in order to meet their daily requirements. This causes environmental issues as it releases large amounts of phosphorus into the areas surrounding the livestock production. One practical and effective method is supplementation diets that are rich in microbial phytases which help enhance monogastric animals' ability to digest phytate.

Feeds that originate from plants are typically stored in the form of hexophosphate ester of myoinositol (phytic acid). It accounts for around 80% of the phosphorus content found in seeds and grains. This phytic acid, its salts, and esters bind to proteins and other important minerals like Mg, Zn, Ca, and Fe thus exhibiting anti-nutritional characteristics by making them less available. To allow liberation of phosphorus, in monogastric animals, which remains bound to phytate, several efforts have been made to overcome this problem by adding external sources of microbial phytases in the animal feeds.

Phytases are produced by a wide range of microorganisms including yeast (such as *S. Cerevisiae*) and fungi (like *Aspergillus ficuum).* These organisms produce phytase that is very active and has high thermostability.

Bacterial phytases have a pH optima in the range of neutral to alkaline and are therefore inactive in stomachs of monogastric animals.

More than 200 fungal strains have shown to be producers of phytase. These include the genera *Aspergillus*, *Rhizopus*, *Penicillium*, and *Mucor*. In another survey, 84 fungal strains belonging to 25 species were shown to be phytase producers and among these, aspergilli produced this enzyme maximally. Among all the tested strains, *A. niger* NRRL 3135 was able to produce phytase extracellularly that was the most active. This strain produces two types of phytases; PhA and PhyB; the former has an optimum pH of 5.5 and 2.5 and the latter has an optimum pH of 2.0. Due to these reasons, filamentous fungi like *A. niger* are reliable sources for production of extracellular phytase.

It has been more than 35 years since the use of microbial phytase in feeds for animals that are monogastric. Several researches have been published that show phytase-treated feeds to enhance the bioavailability of proteins and other minerals. Thus, it provides growth and performance of animals better than those fed with supplementation of phosphates.

Supplementation of inorganic phosphorus causes pollution and therefore fungal phytases have a lot of potential as they are able to lower this by at least 50% in Europe alone. The major source of phosphorus pollution in Europe comes from livestock production and therefore it has been focused on for lowering land pollution of phosphorus. Several countries have made it illegal for farmers to produce phosphorus pollution in land above a certain level per unit area.

It has been estimated that globally, animal feeds of up to 550 million tonnes are currently produced with supplementation of phytases. Therefore, utilization of phytase preparations at commercial levels are prominent. One major limitation of phytase use is that these enzymes become inactive by exposure to temperatures above their thermostability range (65 to 95°C), which are employed during the processed of expansion or feed-pelleting. Ideally, phytase preparations need to remain stable when briefly exposed to elevated temperatures before they are fed to monogastric animals. Moreover, to be useful industrially, phytases used must remain active even after exposure to varied pH ranges within the gastrointestinal tract of these animals.

7.6. STRATEGIES FOR PRODUCTION OF ENZYMES

Enzyme production from fungi is typically carried out using submerged fermentation systems even though solid-state fermentation is prevalently used in Asia. In this section, we will focus on the former method, a flow diagram for which is shown in Figure 7.1.

The stages of fermentation from fungi are quite similar. For antibiotic production, batch fermentation on cheap culture media is scaled up similar to enzyme production. However, downstream stages for processing can vary a lot depending on the desired use of enzyme preparation. Generally, intracellular enzymes, such as lactase, require more steps for processing compared to extracellular enzymes. The common industrial enzymes are extracellularly secreted through large batch fermentation systems and therefore, they do not require much downstream processing. This is an advantage as fewer steps for processing will result in reduced losses in enzymatic yield obtained. In such cases, the processing steps involve the removal of only those materials and enzymes that might alter the catalysis of the desired enzyme. Additional purification steps that are unnecessary are generally avoided as they add to the cost of the process in terms of manpower, equipment, and loss of enzyme yield. Consequently, enzyme preparations for industrial use consist of a mixture of fermentation medium with supplements that stabilize the preparation. If needed, this fermentation broth can be changed into powdered form. The spent culture media can be treated using precipitation by salt or ethanol to generate a protein suspension which can be filtered and dried to produce an enzyme preparation that is highly concentrated.

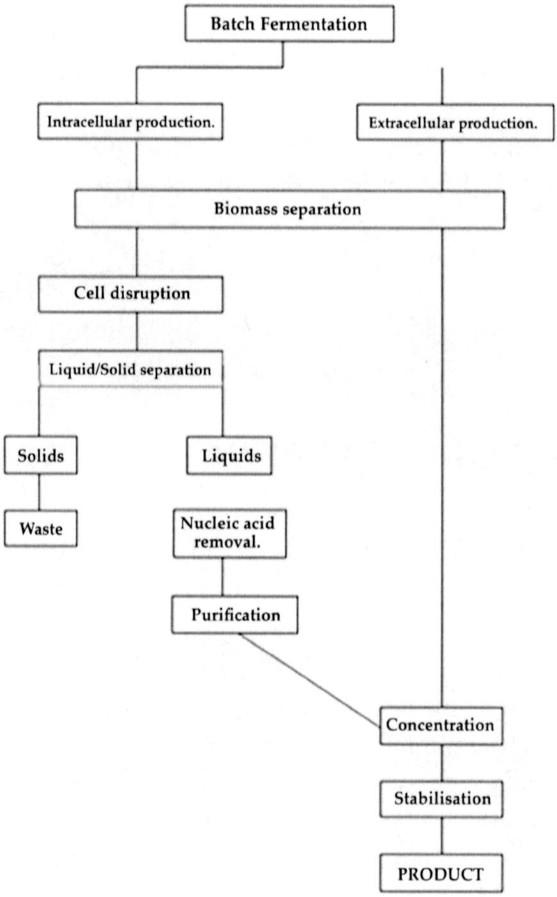

Figure 7.1: A representation of the process used for industrial production of enzymes from fungi. (Source: *Murphy, R. A., & Horgan, K. A. (2005). Antibiotics, Enzymes, and Chemical Commodities from Fungi Fungi (pp. 113–143): John Wiley & Sons, Ltd.*).

7.7. CONCLUSION

Enzymes have been used by mankind since a very long time for multiple purposes. Their power has been harnessed in baking, brewing, making cheese, producing antibiotics and in synthesizing commodities like linen, indigo, and leather. In addition, fungal enzymes also have key roles in industries like paper, detergents, textile, food, and beverages–where they are used for a range of products from wine and fruit juices to tea and coffee.

8
CHAPTER

HETEROLOGOUS PRODUCTION OF FUNGAL PROTEINS IN BIOTECHNOLOGY

CONTENTS

8.1. INTRODUCTION

To express the product of a gene that is normally not present in a cell, we use recombinant DNA technology and we call this a recombinant or heterologous protein expression. Producing recombinant DNA is much simpler compared to expressing recombinant proteins as it involves cutting and joining of segments of DNA in addition to the difficult stages of transcription and translation of the heterologous DNA into a protein that is properly folded and hence functional.

This technique of heterologous protein expression was first developed in the prokaryote E. coli but now it has been extended to numerous prokaryotic and eukaryotic host systems including Gram-positive and -negative bacteria, yeasts, filamentous fungi, insect cells, plants, mammalian cells and transgenic animals.

In fungi, heterologous expression of genes has been attained in various species such as the yeasts *S. Cerevisiae* and *P. pastoris* and the filamentous fungal species like those belonging to the genera *Neurospora*, *Aspergillus*, and *Penicillium.* Proteins produced through this method have been exploited for basic research as well as for commercial uses. In this chapter, we will focus on heterologous protein production in fungal systems for biotechnological uses.

8.2. EXPRESSION OF HETEROLOGOUS PROTEINS IN FUNGAL CELLS

In general, expression of proteins heterologously is done through the following steps; incorporation of heterologous sequences of coding DNA fused with relevant regulatory sequences into specific vectors designed for expression of proteins, this is followed by transformation of the vector into the host cells, and subsequently allowing transcription and translation of the heterologous DNA sequence to produce proteins that are post-translationally modified in order to make them biologically active.

8.2.1. Components Required for Successful Heterologous Protein Expression

8.2.1.1. Heterologous DNA

The heterologous protein that is to be produced can be originated from the genome and therefore requires to have its introns and regulatory sequences processed within its original organism from where it is cloned. Alternatively, the mRNA of interest can be cloned to produce a cDNA sequence of the desired gene through reverse transcription. In this case, it will not have any introns and regulatory sequences within itself. Interestingly, several species of filamentous fungi and yeast have the ability to excise introns from the mRNA transcripts of heterologously expressed genes. Moreover, some filamentous fungal species have the ability to recognize regulatory sequences present in the human genome. Nevertheless, to achieve maximum protein expression, the heterologous mRNA should be transcribed with the help of promoter sequences that belong to genes are strongly expressed within the host cell. Even though production of heterologous proteins is possibly achieved through genomic DNA inserted into fungal genomes, large-scale production is achieved by synthesizing cDNA into an expression vector that consists of promoter and terminator sequences specific to the host cell's gene.

8.2.1.2. Vectors

The bacterial plasmid DNA consists of a characteristic backbone that is feasible for expression into all systems and it also allows manipulation of DNA and purification of plasmid on large-scale in *E. coli.* Expression vectors specific to fungi contain additional features listed below:

1. A strong promoter sequence which allows heterologous mRNA to be produced;
2. A selectable marker appropriate to the fungal host cells;
3. Specific terminator sequences for mRNA transcription and DNA sequences for adding poly-A tail to the mRNA;
4. Appropriate DNA sequences that ensure correct r translation initiation and termination.

Figure 8.1 shows a shuttle vector for expression in *S. Cerevisiae* which is capable of autonomous replication as it has relevant sequences belonging to yeast 2μ DNA – a native yeast plasmid. On the contrary, the vectors that are designed for being functional in methylotrophic yeast *P. pastoris* and in filamentous fungal species like *Aspergillus* and *Penicillium* are called as integrative vectors(shown in Figure 8.2a) and these vectors need to be inserted into chromosomal DNA of the host cell (Figure 8.2b).

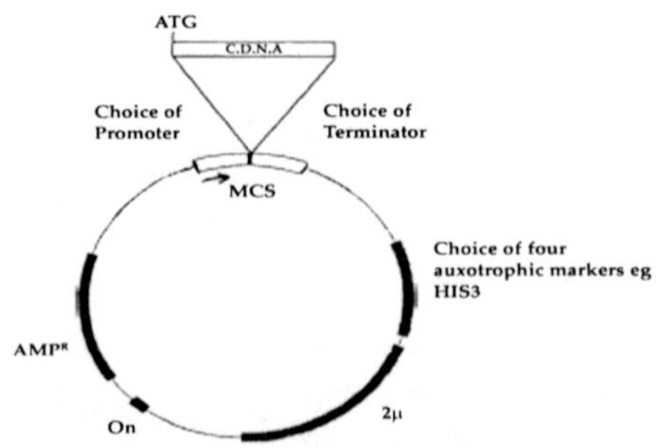

Figure 8.1: The general structure of the expression vector used in *S. Cerevisiae*. (Source: *Curran, B., & Bugeja, V. (2005). The Biotechnological Exploitation of Heterologous Protein Production in Fungi Fungi (pp. 145–169): John Wiley & Sons, Ltd.*).

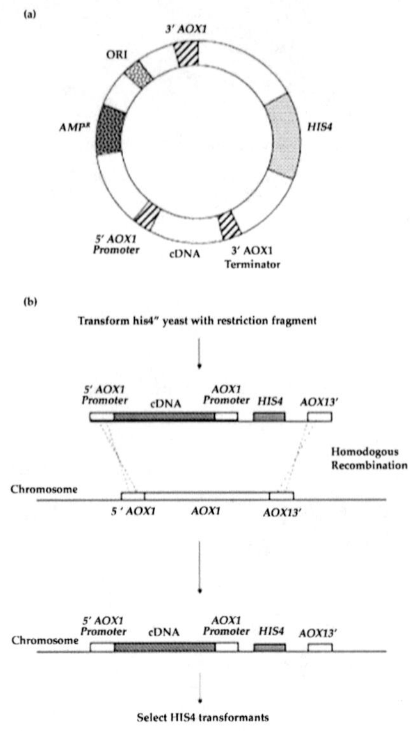

Figure 8.2: a) Structure of *P. pastoris* expression vector b) Its integration into host chromosome. (Source: *Curran, B., & Bugeja, V. (2005). The Biotechnological Exploitation of Heterologous Protein Production in Fungi Fungi (pp. 145–169): John Wiley & Sons, Ltd.*).

8.2.1.3. Transformation and Selection

E. coli can serve as a host for easy manipulation of DNA but to allow its expression into proteins would need the transfer of heterologous DNA to construct into host fungi. The isolated constructs are thus transformed into host-fungal cells and then selected on agar plates using appropriate selection markers. Even though several dominant selectable markers are present which confer resistance to antibiotics in the transformed cells, for selection of positive transformant in filamentous and yeast systems usually require auxotrophic marker complementation. In these cases, the host cells have auxotrophic alleles (e.g., His^{3-}, Trp^{1-}, Leu^{2-}, etc.) which are complemented by wild-type alleles present in the plasmid. Several methods are employed for allowing fungal transformations. These involve enzymatic removal of cell wall and subsequent exposure of the resultant sphaeroplasts to the heterologous DNA in a medium rich in polyethylene glycol and calcium ions, electroporation of cells of yeast and spheroplasts of fungi; transformation by treatment with alkaline cations (e.g., Lithium) of cells of yeast, a process that is similar to *E. coli* transformation.

8.2.1.4. Host Systems

Depending on the type of protein that has to be expressed, the fungal host system is selected. Usually, there are two types of proteins that are most commonly synthesized using biotechnological systems; enzymes and therapeutic proteins. The former has various applications at the industrial level, are needed in large amounts and have low unit value. The latter has high unit value and is needed in low amounts.

To produce enzymes at high amounts for industrial purposes, filamentous fungal systems are ideally used as hosts, such as those belonging to *Aspergillus* species. Many enzymes are produced this way, for example, amylases for fermentation of cornstarch into ethanol, lipases, and proteases employed in detergents, cellulases needed for the production of paper etc. The expression of these heterologous enzymes requires several fungal host species that can help produce enzyme levels at up to grams/ liter quantities (Table 8.1). To produce heterologous proteins that are originated from animals or humans, expression is inefficient in filamentous fungal hosts. This is due to several shortcomings like contamination by protease and low secretory titers of these products.

Consequently, it is preferable to use yeasts like *S. Cerevisiae*, as hosts for heterologous proteins production that have therapeutic significance (Table 8.2). In this section, we will focus on the two biotechnologically significant host systems; the conventional yeast *S. Cerevisiae* and the newly developed *P. pastoris*, which is now becoming increasingly popular. These yeasts are more appropriate for easy genetic manipulations and recovery of products compared to their filamentous cousins. These species, therefore, serve as hosts for heterologous protein production that are commercially significant.

Table 8.1: Some Heterologous Proteins that are Produced by Filamentous Fungi

Host	**Product**	**Origin**	**Use industrially**
A. niger	Glucoamylase	*Aspergillus* sp.	Clarifying fruit juices
A. oryzae	Lipase	*Candida* sp., *Fusarium* sp., *Rhizomucor* sp., *Thermomyces* sp.	Food, textile, detergent, leather, pulp, and paper
A. oryzae, *T. reesei*, *T. longibrachiatum*	Cellulase	*Trichoderma* sp.	Textiles and fruit processing
A. niger, *A. oryzae*	Protease	*Aspergillus* sp., *Rhizomucor* sp. or calf stomach,	
		Aspergillus sp.	Food and leather
	Glucose oxidase		
A. niger			Stabilizing food and beverages

(Source: *Curran, B., & Bugeja, V. (2005). The Biotechnological Exploitation of Heterologous Protein Production in Fungi Fungi (pp. 145–169): John Wiley & Sons, Ltd.*)

Table 8.2: Some Heterologous Proteins that are Produced by Two Yeast Host Systems

Host	Product	Origin	Use therapeutically
S.cerevisiae *S.cerevisiae* *S.cerevisiae* *P. pastoris* *P. pastoris*	Human pap- ilomavirus vaccine Glucagon hepatitis B vaccine hepatitis B vaccine Ecallantide	Human pap- illomavirus Human cDNA Hepatitis virus Hepatitis virus Human cDNA	Vaccine to protect against pap- illomavirus infection Hormone involved in blood glucose level regulation Vaccine to protect against hepatitis B infection Vaccine to protect against hepatitis B infection A 60 amino-acid protein inhibitor of plasma Kallikrein

(Source: *Curran, B., & Bugeja, V. (2005). The Biotechnological Exploitation of Heterologous Protein Production in Fungi Fungi (pp. 145–169): John Wiley & Sons, Ltd.*)

8.2.2. Saccharomyces Cerevisiae

Many useful characteristics of *E. coli* which made it useful as a host for recombinant DNA technology were also found in *S. Cerevisiae*, and therefore, it became the first eukaryotic cell that was used for expression of heterologous proteins. Some useful features of *S. Cerevisiae* include; its rapid cell division, its autonomous replication plasmid, its ability to be transformed without damage to cells, its ability to produce individual colonies on a relatively simple media. Moreover, *S. Cerevisiae* is able to modify heterologous proteins post-translationally, a feature that is absent in *E. coli*. Additionally, these yeast cells are able to produce a few proteins into the medium which can help purify heterologous proteins (a phenomenon which will be discussed later). Contrary to *E. coli*, *S. Cerevisiae* does not make harmful endotoxins or pyrogens and therefore, these cells have long been known to be safe for commercial fermentation processes. These characteristics make it suitable as host systems for production of heterologous proteins that are medically significant in a safe manner.

Even though the yeast expression hosts are versatile, their exploitation for production of heterologous proteins that are bioactive, is still under study. Several factors influence the success of obtaining functional heterologous proteins at acceptable levels, such as, type of expression vector used, the intercellular location of the expressed protein and the protein type.

8.2.2.1. Expression Vector

Many cloning vectors are used in *S. Cerevisiae* cells. In this section, we will focus on vectors that are self-replicating and are therefore commonly used for biotechnological applications (figure 8.1). Vectors of this type, also referred as the YEp (yeast episomal plasmid) vectors are made up of a backbone that has the ability to replicate autonomously. This sequence has been obtained from the yeast 2μ plasmid that is present endogenously. Usually, there are 20 to 200 copies of these plasmids present in each and under favorable conditions, they are present in 60 – 95% of the cell population. In addition, integrative vectors (figure 8.2a and b) are also employed in *S. Cerevisiae*. One such variant of this plasmid functions through homologous recombination to insert several copies of the gene into middle-repetitive 2μ DNA sequences which are formed by the help of the transposable element, Ty. Nevertheless, for commercial applications, expression vectors derived from plasmids are often used.

8.2.2.2. Control of Expression Levels of Heterologous mRNA in Host Cells

Overall, the expression level of heterologous mRNA inside host cell is dependent of the copy number of the expression vector used, the strength of the promoter, and the half-life of the mRNA sequence.

Even though YEp expression vectors have high copy numbers, they are stably inherited only under specific favorable conditions. Production of high levels of mRNA is also controlled by the promoter that is selected for its expression (Table 8.3). The most commonly used promoter is from glycolytic enzymes genes which are expressed at high levels. Production of mRNA constitutively and at high levels is successfully achieved by using promoters belonging to glyceraldehyde-3-phosphate dehydrogenase (*GAPDH*), alcohol dehydrogenase 1 (*ADH1) and* phosphoglycerate kinase

(PGK). Sometimes the heterologous protein can be toxic to the cells under some circumstances. In this case, its expression is controlled by using a regulatable promoter which allows expression to be induced after cells have grown to sufficient amounts. Several regulatable promoters are currently in use, including the one from galactokinase gene (*GAL1*) whose expression is induced more than 1000 times in the presence of glucose compared to galactose. Other than the selection of appropriate promoters, it is also necessary to include relevant terminator sequence. If not, it would result in transcription of mRNA sequences that are abnormally long due to read-through and are unstable. Generally, efficient termination of mRNA is ensured by including the 3' terminator sequence of yeast genes such as *PGK, ADH1*and *CYC1.*

Table 8.3: List of Promoters used for Induction Heterologous Proteins in *S. Cerevisiae*

Promoter	**Strength**	**Regulation**	**Example het-erologous gene expressed using promoter**
PGK, 3-phosphoglycerate kinase	++++	Constitutive	Human b-inter-feron
	+++	Constitutive	
ADH1, alcohol dehydroge-nase 1	++++	Constitutive	Human b-inter-feron
GAPDH, glyceraldehyde-3-phosphate dehydrogenase	+++	1000× induction	Human epidermal growth factor
GAL1, galactokinase			Calf chymosin by galactose
	++	500× repression	hepatitis B surface antigen by phos-phate
PHO5, alkaline phospha-tase			
	+	20× induction by copper	Mouse IG Kappa chain
CUP1, copperthionein			

(Source: *Curran, B., & Bugeja, V. (2005). The Biotechnological Exploitation of Heterologous Protein Production in Fungi Fungi (pp. 145–169): John Wiley & Sons, Ltd.*)

8.2.2.3. Protein Expression at High-Levels

Expression of heterologous proteins depends on the efficiency of mRNA translations and the stability of protein after they are produced. When the expression vector is chosen, it is necessary to consider the regulation of translation of protein and is subsequent translocation. In yeast, the translation initiation phase occurring in mammals is absent. It is, therefore, necessary that translation initiation takes place effectively. This is done by producing cDNA molecules which do not have dyad symmetric region and non-coding AUG codons present within the leader sequences of heterologous mRNAs upstream of the AUG encoding the first methionine in the protein.

In *S. Cerevisiae,* some proteins are formed in soluble forms while others form insoluble complexes. Moreover, some proteins are produced as complexes that are denatured intracellularly. Once harvested, these proteins can be disaggregated and renatured. Using this technology, the first recombinant DNA produced was the vaccine for hepatitis B. some heterologous proteins have smaller half-life due to action of the ubiquitin degradative pathway in the cell and vacuolar proteases. This is especially evident when the cells are broken down for purification. To overcome this problem, detailed information about genetics and biochemistry of *S. Cerevisiae* cells can be exploited to provide powerful tools.

If we use host cells that lack proteases can help enhance the heterologous protein yield and quality obtained. In one commonly used mutant strain (PEP4–3), the inactive zymogen proteases become activated while it is normally inactive. When this protease in inactive, several proteinase activities are inhibited. Interestingly, studying the secretory pathways in yeast cells has made it possible to exploit genetic engineering in order to allow secretion of heterologous proteins out of the cells before they are degraded by any of the above-mentioned mechanisms. This secretion method not only protects the heterologous proteins from the action of proteases, but it also enables the proteins to be recovered and purified easily because there are only a few proteins that are secreted by *S. Cerevisiae*. Two of the handful secretory proteins (invertase and acid phosphatase) are located in the periplasmic space between cell wall and cell membrane and the other two (αfactor and killer toxin) are released out of the cell wall into the surrounding environment (Table 8.4). For the proteins to enter the secretory pathway, they need to have a short hydrophobic N-terminal sequence. Through genetic engineering, these signal peptides can be fused to the DNA sequence of interest for production of secretory heterologous proteins that exit the cell after synthesis. Signal peptides present in all four secretory proteins of *S. Cerevisiae's* have been used in this way, with each showing different degrees of success. Figure 8.3 shows a vector that contains the signal peptide from the gene for α-factor. This type of signal peptide has been used for production of several medically significant proteins like interleukin 2, interferon, and insulin.

Table 8.4: Signal Sequences that Help in Secretion of Heterologous Proteins in *S. Cerevisiae*

Signal sequence	Cellular location of gene product	Example secreted heterologous protein
Invertase	Periplasm	α-1-Antitrypsin
Acid phosphatase	Periplasm	β-Interferon
α-Factor mating pheromone	Culture medium	Epidermal growth factor
Killer Toxin	Culture medium	Cellulase

(Source: Curran, B., & Bugeja, V. (2005). The Biotechnological Exploitation of Heterologous Protein Production in Fungi *Fungi* (pp. 145–169): John Wiley & Sons, Ltd.)

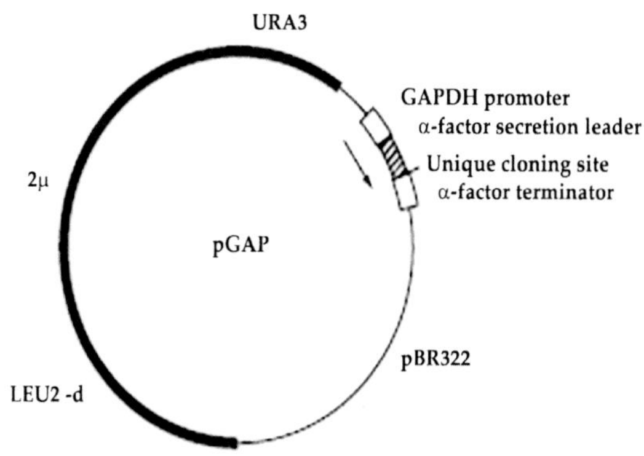

Figure 8.3: Vector that is used for studying secretory proteins in *S. Cerevisiae*. (Source: *Curran, B., & Bugeja, V. (2005). The Biotechnological Exploitation of Heterologous Protein Production in Fungi Fungi (pp. 145–169): John Wiley & Sons, Ltd.*).

8.2.2.4. Proper Folding and Functioning of Proteins

At commercial level, high rate of transcription and translation alone are not sufficient for their applicability. It is necessary to ensure that the proteins that are produced have appropriate structures, are properly folded, and therefore are biologically functional molecules. In mammalian cells, many of the therapeutically significant proteins undergo post-translational modifications. These involve addition or removal of molecules ranging in size from simple chemical moieties like methionine, acetyl group, lipid molecules to more complex sugar molecules which are added into proteins as they are synthesized in cell and are subsequently passed across the endoplasmic reticulum (ER) and Golgi apparatus which adds glycosyl moieties on to them. Many of these complex processes are not seen in simple prokaryotic expression systems like *E. coli*. Therefore, it is more favorable to use eukaryotic systems. Some of these problems are overcome by using *S. Cerevisiae*. For example, production of acetylated superoxide dismutase belonging to humans was failed in *E. coli*. When *S. Cerevisiae* was used, however, it resulted in formation of an identical protein that was soluble, active, and complete with N-terminal alanine.

The secretory system of yeast, moreover, helps ensure proper folding of proteins, removal of methionines at the N-terminal ends and addition of sugar resides to glycoproteins. Intracellular synthesis and extracellular secretion of proteins were compared in case of human serum albumin and prochymosin. When expressed intracellularly, the proteins were insoluble and inactive while secreted forms of proteins were properly folded, soluble, and highly active (Figure 8.4).

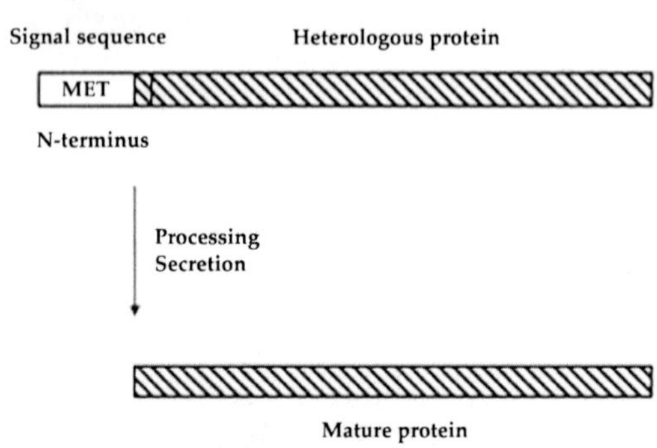

Figure 8.4: Removal of secretory signal peptide from a heterologous protein. (Source: *Curran, B., & Bugeja, V. (2005). The Biotechnological Exploitation of Heterologous Protein Production in Fungi Fungi (pp. 145–169): John Wiley & Sons, Ltd.*).

Sometimes, secretion can be exploited to synthesize proteins that consist of an amino acid other than methionine at their N-terminal end. This is achieve by adding the secretory signal onto the heterologous here at an appropriate amino acid and as a result, the N-terminal methionine responsible for initiation of translation will be present on the secretory signal. In the ER, the cleavage of the heterologous protein will result in formation of a different amino acid at the N-terminal. Similar to mammalian cells, yeast cells are also able to carry out both types of glycosylation of proteins; N-linked and O-linked on the amino acid sequences Asn-X-Ser/Thr and Thr/Ser, respectively. The glycosylated proteins in yeast are attached to different outer core carbohydrates that also vary in number compared to those present on proteins synthesized in mammals. As result, the proteins expressed in yeast are likely to cause immunogenic reactions if they are used for therapeutic purposes. This problem can be overcome by exploiting site-directed mutagenesis to change single amino acid within *E. coli* to change the sequence of glycosylation recognition site before the protein is expressed in *S. Cerevisiae.* This method has been applied for production of urokinase-type plasminogen activator.

8.2.2.5. Drawbacks

Even though *S.cerevisiae* is able to produce a wide range of proteins, it poses some limitations in the process. In some cases, the heterologous proteins become hyper-glycosylated due to presence of a very primitive glycosylation system. Moreover, metabolism of glucose results in formation of alcohol which reduced biomass formation and consequently production of heterologous protein is limited. Nevertheless, *P. pastoris* is another yeast system that has been developed which grows to higher densities and does not carry out hyper-glycosylation of its protein. Currently, it is less prevalently used for production of therapeutic proteins compared to *S. Cerevisiae*, however, with developments in glycobiology, it is expected this species will have increased applicability at commercial level (Table 8.5).

Table 8.5: Yeast Origin Heterologous Proteins that Have Therapeutic Significance

Period	Fungal hosts	Number of products	Types of product
1985–1990	*S.cerevisiae*	1 (1986)	First recombinant Hepatitis B vaccine
1991–1995	*S.cerevisiae*	1	
1996–2000	*S.cerevisiae*	10	Growth factor for stimulating bone marrow
	P. pastoris	1	5 vaccines, an anticoagulant, a tissue growth factor, 3 hormones
2001–2006	*S.cerevisiae P. pastoris* *S.cerevisiae* *P. pastoris*	8 2 4 1	hepatitis B vaccine
2006–2010			5 vaccines, 2 hormones, 1 therapeutic enzyme Insulin and interferon 3 hormones, 1 vaccine Ecallantide, a protein inhibitor of plasma kallikrein

(Source: *Curran, B., & Bugeja, V. (2005). The Biotechnological Exploitation of Heterologous Protein Production in Fungi Fungi (pp. 145–169): John Wiley & Sons, Ltd.*)

8.2.3. Pichia Pastoris

Pichia pastoris includes some species of yeast which come under the*Torulopsis*, *Pichia, Hansenula, and Candida* genera. They are methylotrophs which means that they can use methanol as their source of carbon through specific metabolic pathways. The promoters present upstream of these genes are very strong, very sensitive to methanol and express the genes at high levels in its presence. This makes them suitable for use in expression of heterologous proteins. Like *S. Cerevisiae,* these species are also easily manipulated genetically. In addition, they offer several advantages over it:

1. They are compatible with integrative vectors which eliminates the requirement for selection media within the fermenters and at the same time exhibits high mitotic stabilities within the recombinant strains;
2. Because they don't produce any toxic ethanol, these species are able to grow in the fermenters up to much greater cell densities;

3. The patterns for heterologous protein glycosylation are more authentic.

Among all the available species, *P. pastoris* is the most prevalently used species due to its ability to produce products that are highly valuable and have therapeutic significance. *P. pastoris* system has a commercial kit available and therefore has been widely used to expression of more than 100 individual proteins which are involved in basic research and for industrial uses. Even though several dominant selectable markers are also used, to select transformants that express the heterologous gene, an auxotrophic complementation marker *his4* is commonly used. Unlike *S. Cerevisiae* which uses autonomously expressing vectors, these species employ integrative vectors. Within an *E. coli* vector, the *AOX1* gene which is present between the promoter and terminator sequence is replaced by the gene of interest through splicing. The vector also consists of *His4+* gene and further downstream of this is the sequence of 3' end of *AOX1* gene (Figure 8.2). A linear sequence of gene of interest which is flanked by AOX1 sequences are transformed into host cells that are *His4-*. When inside the host, this gene construct can be inserted into the locus of the chromosome which consists of AOX1 gene through homologous recombination (Figure 8.2b). These transformed cells can be grown on glucose and is repressed until it is exposed to methanol which induces its expression. Alternatively, the cells can be grown on methanol by using locus for alcohol oxidase which allows constitutive expression of heterologous protein. The former allows very tight regulation of heterologous expression of genes. Some available integrative vectors allow targeting of the construct into locus consisting of *His4* gene. Therefore, this easily regulatable promoter offers several practical advantages over the galactose-regulated promoters used for protein expression in *S. Cerevisiae*. Moreover, *P. pastoris* is also able to glycosylate the proteins more efficiently and faithfully. The signal sequences that are most frequently used are a-factor pre-prosequence isolated from S. cerevisisae and the *P. pastoris'* gene for an acid phosphatase.

Pichia has helped produce a number of therapeutically important proteins (Table 8.5). it is expected that this number is going to increase sharply because recent genetic manipulations of *P. pastoris* has enabled production of glycoproteins that are exactly similar to those produced in humans in every aspect. This was achieved by genetically knocking out glycosylation genes inherently present inside the yeast's genome and inserting 14 glycosylation genes present in humans in order to produce a recombinant strain which is able to synthesize proteins with uniform glycosylation and have a capping of sialic acid. For therapeutic proteins, it is necessary that they harbor specific sugar molecules to allow them to be biologically active and to prevent undesirable immunogenic reactions. Earlier it was not possible to produce heterologous proteins in yeast cells which consist of the correct pattern of sugar molecules but recently, *P. pastoris* has been modified to allow it to do so and in fact, their ability to N-glycosylate heterologous proteins is even more efficient compared to the mammalian cells that are employed for production of heterologous proteins that are specifically targeted for glycosylation.

Since the production of the first heterologous protein in yeast cells, the vaccine for hepatitis B in 1986, the yeast host cells have seen many developments to serve for heterologous protein production.

8.3. A CASE STUDY FOR PRODUCTION OF VACCINE FOR HEPATITIS B: A YEAST PROTEIN PRODUCED HETEROLOGOUSLY WORTH A BILLION DOLLARS

The annual market of Hepatitis B sub-unit vaccine is up to US$ 2 billion. It is one the biggest examples of success in molecular technology. This protein is produced heterologously in *S. Cerevisiae* and more recently in *P. pastoris* and *H. polymorpha*. It is remarkably significant example that illustrates the success and development of biotechnology for production of important heterologous proteins.

8.3.1. Hepatitis Disease

Hepatitis B virus is double-shelled and it belongs to the class Hepadnaviridae (Figure 8.5). It causes more than 250,000 human deaths annually throughout the world. They can cause viral infection in the liver in two ways; either acutely or chronically. In acute hepatitis, the disappearance of symptoms is correlated to clearance of virus from the body. On the other hand, getting rid of symptoms in chronic hepatitis does not necessarily mean that the viral titer in the body has been cleared. The virus stays in the body and the patient becomes a carrier of the disease. The chronic hepatitis is often associated with primary hepatocellular carcinoma and progressive cirrhosis. An estimated number of men, women, and children carrying this disease is around 200 million worldwide.

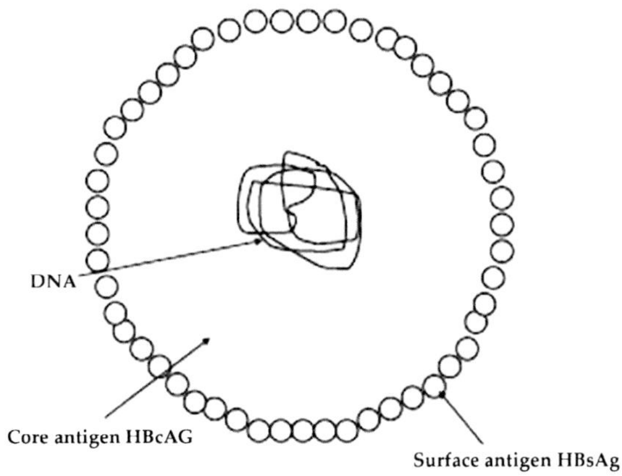

Figure 8.5: Hepatitis-B virus. (Source*: Curran, B., & Bugeja, V. (2005). The Biotechnological Exploitation of Heterologous Protein Production in Fungi Fungi (pp. 145–169): John Wiley & Sons, Ltd.*).

In most of the cases, the antiviral interferon therapy is not available or when available is ineffective because the symptoms have already appeared and developed. Therefore, with respect to this disease, vaccination serves as the only useful medical intervention. Or in other words, there is a huge market opportunity for vaccines produced against Hepatitis B. This is the reason so many companies and organizations are working on production and sale of these products.

8.3.2. Production of Vaccine from Hepatitis-B Carrier Patients

The vaccine for hepatitis B was commercially made for the first time using viral protein particles isolated from human plasma of chronic hepatitis carriers. These sub-viral viral glycoproteins, consisting of the hepatitis B surface antigen (HBsAg) located within the phospholipid layer, are 22 nm in size and are non-infective products but are extremely immunogenic. When these non-infective protein particles are exposed to our immune system, it considers it as the complete 42 nm viral proteins. Therefore, it elicits a strong immune reaction against the viral coat and hence helps in conferring resistance to future attacks by the actual virus.

In 1981, it was licensed and in USA, it was successfully used thereafter for more than 10 years. It proved to be effective, well-tolerated, and safe. However, it required a continuous supply of plasma and therefore needed a lot of processing and testing for safety. Even though this vaccine was expensive to synthesize, there was unbased fear among the users who expected it to induce contamination, due to its potential to exhibit infective properties of hepatitis B virus or due to transfer of other diseases transferred via donor's blood.

The problems associated with this vaccine derived from humans were, therefore, overcome using recombinant DNA technology. During the early 1980s, the techniques for genetic manipulations were not very well-developed and to express a heterologous human protein in a system like *E. coli* required extensive efforts. For hepatitis B vaccine to be effective and useful, it had to be expressed sufficiently. To achieve this, the biotechnologists had to isolate the coding DNA sequence for HBsAg protein from the complete genome of the virus. Subsequently, it had to transformed into the host cells and allowed to transcribe and translate into a protein that was safe and immunogenic at the same time.

8.3.3. Recombinant Vaccine production using Genetic Engineering

The hepatitis genome with a size of 3200 bases was already known. From this genomic sequence, DNA sequence encoding for the majority of the surface proteins consisting of 226 amino acids was isolated. Subsequently, it was spliced into an expression vector specific for E. coli downstream of a strong promoter and eventually transformed into specific host cells. A viral protein was expressed successfully but it non-glycosylated. Unfortunately, the host cells were unable to synthesize the complete 22 nm phospholipid-protein particles. In the absence of the complete and assembled protein particles of HBsAg, the immunogenicity elicited was 1000 weaker compared to that induced by the plasma-derived vaccine. Therefore, it was not surprising for the researchers to observe failure of this recombinant protein in inducing an immune reaction in animal body.

Staying focused on their target, the scientists then attempted to use recently developed technique of gene cloning in the field of recombinant DNA technology in *S. Cerevisiae*, a eukaryotic organism. It was expected even though the prokaryotic E. coli failed to produce properly glycosylated secretory proteins but the eukaryotic *S. Cerevisiae* system would enable this and then this protein will be immunogenic. For meeting this target, the scientists inserted the DNA sequence coding for HBsAg in the shuttle vector based on E. coli but having the replication origin of 2μ and the TRP1 gene. The gene was placed under the control of promoter for the yeast alcohol dehydrogenase (ADH1) (Figure 8.6a). This construct was then transformed into a trp1- host and the positive transformant selected and observed to express the HBsAg proteins at sufficient levels. Similar to the proteins isolated from carrier human plasma, these expressed proteins also aggregated into phospholipid particles. The only feature present in the former isolated protein which was absent in the yeast-derived protein was glycosylation and ability to be secreted extracellularly. Instead, this protein was unglycosylated and formed aggregates inside the cells. Interestingly, even though the protein lacked glycosylation, it was still able to elicit an immunogenic response in the tested animals. This indicated that glycosylation was too necessary for exhibiting immunogenicity and for assembling of the protein particles.

8.3.4. Scaling up the Production of the Produced Vaccine

The expression vector was improved (Figure 8.6 b) by using the basic expression vector which could produce 22 nm phospholipid-protein particles that were immunogenic and helped overcome the problems associated with the vaccine isolated from the human carrier plasma. This improved version was further developed to scale-up the production of vaccine to industrial level. Several changes in the vector used in principal experiment were added to produce a vector that was applicable at industrial level to enable production of heterologous protein at high levels in yeast.

Both plasmids are shuttle vectors which have common features; they are derived from the yeast 2μ plasmid and also harbor DNA segments belonging to plasmid isolated from *E. coli*. The vector used for the proof of principle experiments consisted of a *TRP1* marker, while the newly developed industrial vector consists of *LEU-2d* marker. The latter gene, however, has a promoter that is truncated and is therefore not transcribed effectively. Therefore, to enable the recombined strains to propagate in the absence of leucine, it is necessary to insert the plasmid at high copy number. Ideally, the cells should be having 150 to 300 copies of plasmid consisting of *LEU-2d* gene per cell. This is in contrast of just 30 copies of the *TRP1*-containing plasmid.

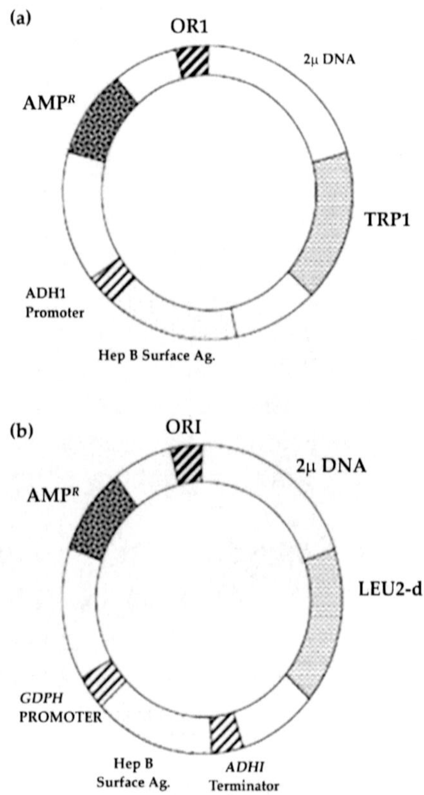

Figure 8.6: Plasmid used for a) proof of principle experiments and b) industrial scale-up. (Source: *Curran, B., & Bugeja, V. (2005). The Biotechnological Exploitation of Heterologous Protein Production in Fungi Fungi (pp. 145–169): John Wiley & Sons, Ltd.*).

Moreover, to ensure efficient termination of transcription, the industrial vector harbors a terminator sequences which is absent in the proof of principle vector. The reasonably strong promoter of *ADH1* gene in the proof of principle vector is replaced by a well-characterized and more efficient promoter of the *GPDH* gene in the industrial sector.

Conclusively, other than the basic expression sequences, the industrial vector compared to the proof of principle vector has additional sequences which ensure its high copy number, a more effective initiation and termination of the transcription. Resultantly, these alterations ensure production of mRNA at high levels and thus the expression level of heterologous protein is also enhanced significantly within the host cells harboring the industrial strain compared to the proof of principle host cells.

Production of hepatitis B vaccine in *S. Cerevisiae* not only helped in successful production of proteins at high levels, it also helped elucidate some mechanisms involved in expression of heterologous protein and regulation of gene. These findings proved to be a key to biotechnological developments till date. It was also the first successful exploitation of a eukaryotic host for commercial production of a heterologous gene. With this began the novel era of development of vaccines. Beyond the developments in expression technology, this success helped protect millions of lives which were risked against Hepatitis B, the deadly, dangerous disease.

8.4. EXPRESSION TECHNOLOGY AND ITS FURTHER USES IN BIOTECHNOLOGY

Protein expression technology in *S. Cerevisiae* can be used not only for commercial protein proteins but it can also be exploited as an *in vitro* tool for studying molecular interactions. These systems used for expression of heterologous proteins can be altered in aspect of their promoters, transcription factors and signal cascades in order to elucidate interactions of heterologous proteins at molecular levels and can be scored easily based on the phenotypes observed. Screening for molecules that interact and alter these interactions can be carried out by biotechnologists through employment of these assays. Due to this, these molecules are therefore called as the lead molecules for development of drugs. The expression of heterologous peptide receptor and steroid proteins has been examined using marker genes and selection media, respectively. Moreover, the protein interactions intracellularly can be studied by using the elegant yeast-two hybrid technique which exploits the linked nature of transcription factors.

8.4.1. Heterologous Receptor Proteins, Their Expression, and Analysis

Breast cancer has shown to be associated with a key hormone in humans, estrogen. Similar to other steroid hormones, estrogen also functions by influencing expression of genes by physically binding to receptor proteins that are present at the cytoplasmic regions. Once formed, this complex then moves into the nucleus of the cells and binds to estrogen receptor element (ERE), which are DNA sequences of promoters having specific recognition sites for the receptor. These receptors or ERE sequences are generally absent in yeast cells. However, it has been possible to genetically engineer the yeast cells by inserting at chromosomal locus of *URA3* gene, the gene sequence of β-galactosidase gene which is under the control of the inactivated promoter of *CYC1* gene having sequences for the ERE (Figure 8.7a). Simultaneously, these cells also contain plasmids that express the intracellular receptor protein constitutively (Figure 8.7b). Therefore, when these genetically engineered cells are grown and exposed to estrogen, it binds to the receptor and goes into the nucleus. It recognizes the ERE sequence present within the promoter present upstream of β-galactosidase genes resulting in an induced expression. Assays have been developed for expression of this enzyme.

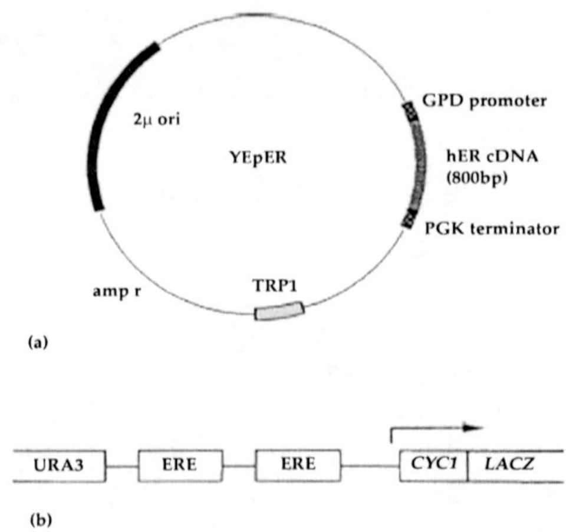

Figure 8.7: Estrogen receptor protein analysis. (Source: *Curran, B., & Bugeja, V. (2005). The Biotechnological Exploitation of Heterologous Protein Production in Fungi Fungi (pp. 145–169): John Wiley & Sons, Ltd.*).

The activity of the receptor induced by estrogen is then directly proportional to the amount of enzyme that is expressed. This enables comparing different tested cells and identifying their ability to inhibit the activity of estrogen. This system has been successfully exploited for detection of effects induced by site-specific mutations on the binding efficiencies of the hormones and subsequently measuring their abilities to function as either the agonists or antagonists of the hormone.

When the steroid hormone receptors are activated by hormones, they directly bind to the promoter elements present within the DNA. On the other hand, the peptide hormones bind to their respective receptor proteins which are embedded within the cell membrane. These interactions at molecular levels are signaled to the kinase cascade present in the nucleus. These kinases are basically enzymes that catalyze addition of phosphate molecules to the proteins. Once activated, this cascade results in phosphorylation of a protein which is transcription factor that subsequently becomes activated and induces expression of genes downstream of specific promoter elements. A class of heterotrimeric protein complexes present in the membrane close to the receptor proteins are called the G-proteins and they control this type of signal transduction. These proteins become active when the GDP molecule which is constitutively bound to them is replaced by a molecule of GTP. As mentioned above, the G-proteins consist of three sub-units, a- which remains in contact with the receptor protein and b and y, which help communicate with the kinase cascade. When a specific peptide binds to the receptor, the GDP molecule bound to itsα sub-unit is replaced by a GTP molecule due to some induced changes in its conformation. As a result, this replacement causes the dissociation of the $\beta$and γ sub-units. The dissociated sub-units translocate towards another protein within the cell membrane which stimulates the kinase cascade, and as a result, specific transcription factors become phosphorylated, activated, and induce expression of relevant genes. Depending on the type of organism under consideration, the G-protein and receptor structures vary and the elements for promoters and transcription factor that are influence also differ. However, the mechanism for signal transduction from the signal at the membrane to an alteration in expression of genes in highly conserved. To study the interaction between human peptides and their respective membrane receptor targets, the *S. Cerevisiae* signal transduction pathway for mating has been exploited (Figure 8.8).

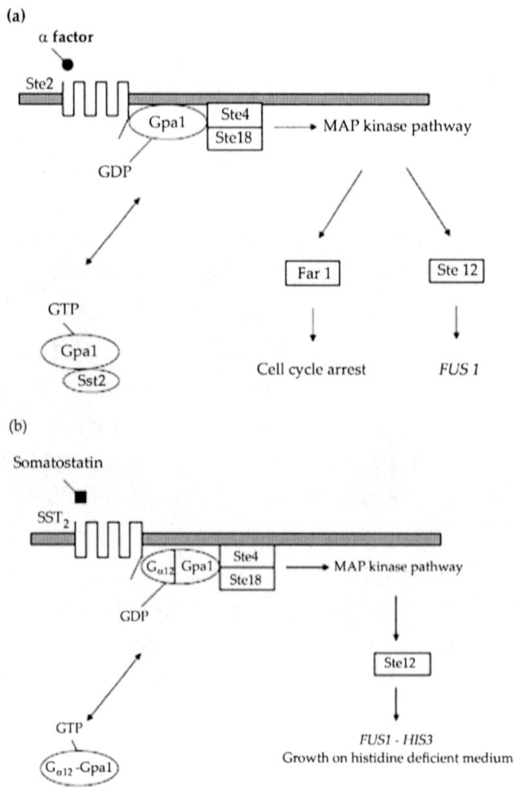

Figure 8.8: Re-engineering of a pathway in yeast for analyzing a heterologous receptor. (Source: *Curran, B., & Bugeja, V. (2005). The Biotechnological Exploitation of Heterologous Protein Production in Fungi Fungi (pp. 145–169): John Wiley & Sons, Ltd.*).

Yeast haploid cells synchronize their cell cycles with their opposite mating cell type to allowing mating by secreting small peptides which combine with the cell membrane receptors of the opposite mate. α cells produce α factor which binds to the receptor protein present in normal haploid cells of the 'a' mating type. This results in activation of a G-protein present close to the receptor in the 'a' mating type. The pattern of gene expression in the cell is altered as result of this kinase cascade. Some of the genes include those involved in arresting the cell cycle in the G1 phase, expression of the gene products prepared for the cell and subsequently fusion of nuclei of the synchronized cells that finally mate. To enable the exploitation of yeast cells for screening using high-throughput technologies, these cells have been altered to exhibit a 'read-out' that indicates selective growth of cells in the medium. Such tools have been successfully developed and used for screening of multiple drugs because the re-engineering of the 'mating pathway of 'a' type mating yeast cells offers several advantages such as:

1. Instead of the α-factor, they are able to express a human receptor protein;
2. It was able to express a genetically-engineered α-subunit of the G-protein which was able to interact with the β and γ subunits present in the yeast and simultaneously with the human receptor protein;
3. The kinase cascade involved in the mating pathway activates a transcription factor which binds to promoter element upstream of the *HIS3* wild-type gene;
4. Once the kinase cascade is activated, the arrest of cell-cycle is prevented due to deletion of the *FAR1* gene.

Because these cells do not have the receptor protein for a-factor, they are no longer responsive to addition of a-factor. Nevertheless, addition of human peptides into the cells results in binding to the heterologous receptor protein present in the membrane (Figure 8.8). This causes conformational changes which are detected by the re-engineered a-subunit of the G-protein. Subsequently, the GDP molecules displace the GTP molecules on it, the b and y subunits dissociate and translocate to the nucleus where they initiate the kinase cascade. The resultant activated transcription factor induces expression of those genes that are normally expressed in the presence of the A-factor. In addition to this, these transcription factors also bind to the promoter elements present upstream of *HIS3* gene. As a result, HIS+ phenotype is conferred to the cell. Because the *FAR1* gene has been deleted in this mutant strain, the cell will no longer be arrested in the G1 phase of the cell cycle. The cells, therefore, grow and divide in the His-selective medium.

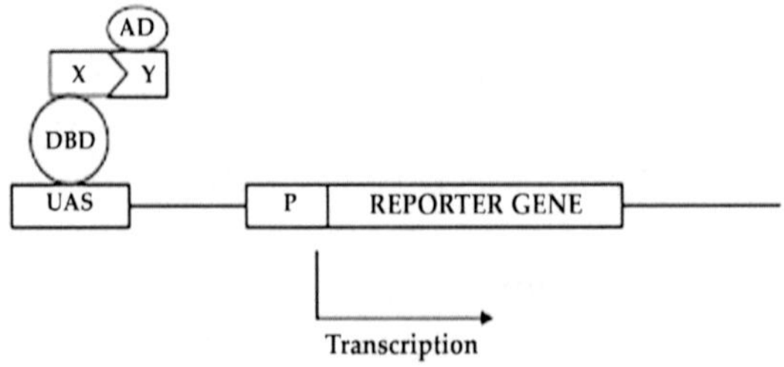

Figure 8.9: The yeast-two-hybrid system. The reporter gene is activated only when 'X' and 'Y' interact with each other. (Source*: Curran, B., & Bugeja, V. (2005). The Biotechnological Exploitation of Heterologous Protein Production in Fungi Fungi John Wiley & Sons, Ltd.*).

8.4.2. Yeast-two Hybrid System: A System that Enables Phenotypes Scoring Based on Protein-Protein Interaction

This efficient system for protein expression works on the basis that the transcription factors are made up of two individual domains: the activation domain and the DNA binding (AD and DBD). The two expression vectors are expressed separately, they are unable to interact and therefore construction of transcription factor that is function is not possible. They are made functional by expressing them fused with two proteins, X and Y, which interact with each other bringing the two domains in close proximity to each other and hence generating a functional transcription factor (Figure 8.9).

Some vectors of yeast have been developed which consist of DBD belonging to the GAL4 transcription factor and the AD of the same protein are present on another expression plasmid which is based of 2μ yeast sequences and also consists of selectable markers. The two genes of interest are fused to the sequences of the DBD and AD, respectively. Haploid cells that exhibit complementary genetic markers are used as host cells for transformation of these plasmids. These host cells have a defected *HIS4* genes and one of the two cells harbor a promoter that is inducible by GAL-4 and which is present upstream of HIS4+ coding sequence. The two strains are allowed to mate with each other. An interaction between the two proteins under study will enable the AD and DBD to come close to each other and hence assemble them to form a complete functional transcription factor. This results in expression of His4 gene and the positive strains are able to survive on a selective His-medium. If the two proteins of interest are not interacting, the two domains of the transcription factor will not assemble together and even if the DBD binds to the promoter of His4 gene, transcription will not be initiated as there will be no AD (Figure 8.9). As a result, the diploid strain will not express *HIS4* and will be unable to grown on His-selective medium. This system, called as the yeast-two-hybrid system has been exploited to study numerous interactions between proteins belonging to various species. In addition, it can also be used to identify small molecules that interfere with specific protein-protein interactions and therefore it serves as a tool for drug discovery.

8.5. CONCLUSION

Fungal heterologous protein expression involves production of heterologous (mostly fungal) proteins by filamentous fungi for commercial applications, production by yeast species proteins that are highly valuable for therapeutic purposes and the exploitation of cells of yeast to detect protein-protein interactions between heterologous proteins originated from various organisms. Production of heterologous proteins initially was an experimental approach where hosts were tested for their ability to produce proteins that were authentic and high yielding, development of next-generation techniques has improved the promising outcomes of this approach. One prominent example currently in the genetic engineering of *P. pastoris* which is now able to synthesize human proteins that are glycosylated and are reproducible in much better way than those produced by mammalian host cells. There will much more to offer in this field. Fungal global transcriptome studies have helped reveal factors that are limiting its rate of it pathways involved in protein folding and secretion. These aspects are, therefore, rate-limiting for heterologous expression of proteins. Since the last two decade, this rational approach has been successfully applicable to fungi, particularly yeast, and it is expected that their role is going to be more important in the coming years for heterologous protein production.

9
CHAPTER

FUNGAL PROTEOMICS

CONTENTS

9.1. INTRODUCTION

The study involving complement proteins of fungi both extracellular and intracellular is referred to as fungal proteomics. Previously protein isolation was carried out through techniques such as chromatography, followed by characterization using enzymes, and the identification of their sequences at the N-terminus or partial amino acid sequence. With advances in genome sequencing techniques and mass spectrometry, along with protein separation techniques at high-resolution (e.g., electrophoresis which separates proteins on the basis of their molecular weight), it has become possible to carry out simultaneous isolations, separation, and identification of hundreds or even thousands of proteins from fungi. Even though this large-scale approach has the limitation of generating a vast amount of data, it serves as a tool for fungal proteomics in the current situation.

There is a constant evolution of investigations and strategies used for fungal proteomic analysis, nevertheless, all of these have some common features:

1. There are databases available which contain genomic sequences or cDNA sequence for the fungal species of interest. Moreover, there may be genomic data available for fungal strains that are related to the fungal genome of interest (e.g., Pubmed)(See Chapter 3);
2. There are optimized protocols to culture fungal cells and to isolate both intracellular and extracellular fungal proteins;
3. Proteins of fungal origin can be separated using techniques like electrophoresis such as sodium dodecyl sulfate-polyacrylamide gel electrophoresis (SDS-PAGE), or chromatography. (Figure 9.1);
4. The ability of proteolytic trypsin to digest individual fungal proteins into smaller peptides;
5. The obtained peptide mixture can be analyzed using liquid chromatography-mass spectrometry (LC-MS). This approach enables the peptides to be separated from one another and hence determine their sequences and masses and using the data obtained for comparison to DNA or cDNA databases;
6. The *in silico* translation of the required proteins is followed by identifying the proteins or protein sets which are coded by

specific gene sequences allowing the researchers to conclude that the peptides obtained must be derived from the protein encoded by this gene.

In this chapter, we will basically focus on the proteome study of filamentous fungi, e.g., *Aspergillus* spp. This is because these fungi consist of invaluable information about proteins products and serve as 'black boxes' in the aspect of information they provide about the biochemical pathways involved in these fungi for survival in the environment. Even though filamentous fungi have a lot to offer in the biotechnology fields like enzymes and antibiotic production, there are some specific reasons that the fungal proteomics is not as prevalently used as bacterial and viral systems. These include:

1. The genomic sequences of filamentous fungi were not available before the 2000s, this implies that the protein MS technique was not applicable in order to attain their protein identification;
2. Contrary to genomes of yeast, virus, and bacteria, the fungal genomes consist of several introns within their genes. There was inadequate information about intron-exon splice regions within genes due to lack of relevant bioinformatic tools. This meant that the data obtained from peptide mass could not be interrogated against the genomic DNA databases already available;
3. Due to the presence of a tough cell wall in fungi along with the presence of several carbohydrate polymers which interfere with the process, it was difficult to isolate proteins in fungi that were expressed intracellularly;
4. In many wild-type organisms, the fungal proteins are not expressed and secreted at sufficient levels and it is greatly dependent on the culture composition and conditions set for growth.

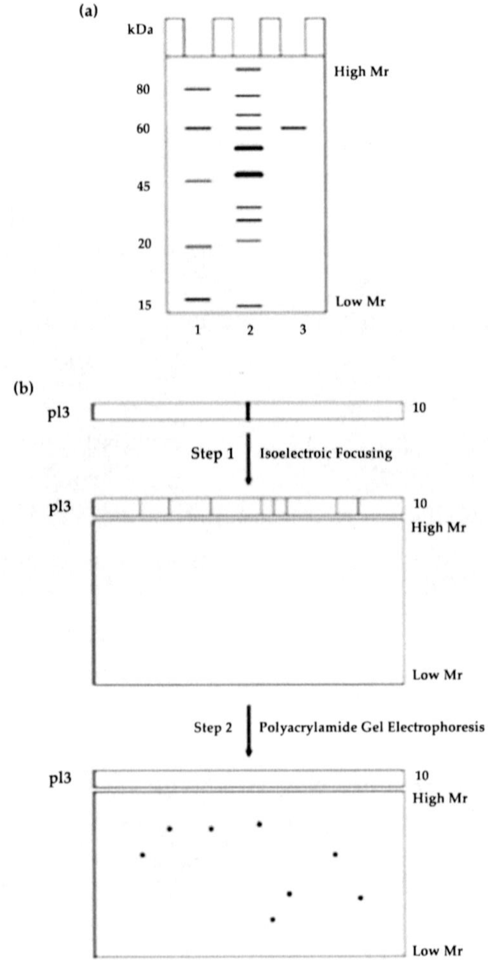

Figure 9.1: Diagrammatic presentation of SDS-PAGE and 2D-PAGE (two-dimensional polyacrylamide gel electrophoresis. A)Proteins are separated on SDS-PAGE on the basis of their molecular weight. Lane 1 contains the marker proteins which consists of proteins with known molecular weights. Lane 2 has extract proteins that have various molecular weights and lane 3 shows a single band which contains a preparation of purified proteins. B) 2D-PAGE involves two major stages; isoelectric focusing (IEF) followed by PAGE. In the latter stage, the proteins are separated on the basis of their charges on an IEF strip. In the former step, the strip obtained at the end of IEF is placed on top of a PAGE gel which allows further separation of proteins on the basis of their molecular weights to produce 'spots' of proteins. These spots can be excised and used

for further analysis through trypsin digestion and subsequent mass spectrometry techniques. (*Source: Sean Doyle (2011). Fungal proteomics: John Wiley & Sons, Ltd.)*.

Therefore, fungal proteins which are significant medically and biotechnologically can be identified rapidly using developed proteomic techniques, making fungal proteomics a newly emerging and significant field. This field can also be employed for studying protein modification and protein-protein interactions. In this chapter, we will discuss various methods used for extraction, isolation, purification (e.g., electrophoresis) and identification (e.g., MS). Along with this, we will discuss specific examples for using proteomics to study the potential of fungi in virulence and commercial purposes.

9.2. ISOLATION AND PURIFICATION OF PROTEINS

9.2.1. Strategies for Lysing Cells

Before the proteins are analyzed, they need to be extracted and isolated from fungi. in animal cells, it is ready to achieve lysis of cell membrane through the use of detergents. However, in fungi, there is an additional barrier of cell wall at the outside of cells which hinders the extraction of fungal intracellular proteins in an efficient way. Therefore, to isolate proteins from fungal mycelia, several vigorous techniques have been developed. These include; sonication, been-beating., hyphal maceration in liquid N2 using mortar and pestle, and rapid pressure changes (French Press technology). One or combination of more of these techniques will allow disruption of fungal mycelia. Several other factors have to be controlled to achieve this and also for obtaining sufficient intracellular protein yields, including, presence of protein inhibitors, strict control of pH and temperature (2–8°C), and optimum ratio of eactraction buffer:mycelia. Usually, lysis of mycelia is followed by centrifugation at high speeds for removal of insoluble substances (e.g., intact mycelia, fragments of cell wall or cell debris) in order to obtain a supernatant that is clear and rich in proteins. Further processing of this supernatant can be carried out using techniques like ammonium sulfate fractionation, dialysis or volume reduction by ultrafiltration, prior to protein purification by chromatographic techniques.

9.2.2. Chromatography

We will not be discussing details of the chromatography technique in this chapter. However, we will be discussing ion-exchange chromatography which allows separation of proteins on the basis of their charge, gel filtration which allows isolation of proteins based on their molecular weights, and affinity chromatography which facilitates isolation of proteins on the basis of their affinity towards specific ligands. The merits and demerits of each of the chromatographic techniques are summarized in table 9.1. Once isolated, the protein is subsequently characterized in various aspects. For example, if the protein is an enzyme, its activity is determined, if it is an antibody, its immunogenicity is tested, and in addition, its interaction with other proteins and its sequence is obtained. Usually, chromatography is used for isolating protein that are prepared for further use. It is not used directly as a tool for analysis of global protein isolation for identification and sequence determinations.

9.2.3. Protein Extraction before 2D-PAGE

2-D PAGE is a protein separation technique which is based on both charges and molecular mass of proteins and helps to quickly (1 to 2 days) yield proteins that are pure and have sufficient concentrations. Subsequently, they can be digested by trypsin to yield shorter peptides which can then be used for further analysis and identification using MS. This is in contrast to chromatography techniques which take days or even weeks to obtain high yields of very pure proteins. Luckily, it is possible to extract mycelial proteins directly into reagents which are then separated using 2D-PAGE. In this approach, the mycelia are first ground in liquid nitrogen and then lysed in solubilization buffers through sonication. These buffers consist of high amounts of urea, thiourea, and detergents, along with ampholytes. Eventually, the extracts of proteins obtained in this approach can be employed directly to isoelectric focusing strips (pH range 4–7 or 3–10) which allows separation of proteins based on their isoelectric point before their separation based on their molecular weight (though protein electrophoresis) (Figure 9.1).

9.2.4. Extraction of Proteins and Fragmentation by Enzymes before Purification

Before protein analysis, it is not always necessary to achieve protein fractionation. It is common to use an approach which involves digestion of complex protein mixtures using trypsin and subsequently identify the yielded peptides using LC-tandem mass spectrometry (MS/MS) technique. The obtained data is then compared to data already available in *in-silico* databases. This overall approach is termed as the 'shotgun proteomics.' These types of analyses utilize the power of computing and signify the limit of usefulness of proteomic techniques to globally identify proteins. To achieve this, the total content of protein is first extracted by grounding mycelia in liquid nitrogen with addition of high concentrations of agents that denature proteins, such as, guanidine-HCl and for reduction, such as dithiothreitol (DTT). As a result of this treatment, all the cellular proteins are solubilized and released. When certain protease inhibitors are added, these extracted proteins are further protected from general protein degradation. Once solubilized protein mixture is obtained, they are treated with alkylation agents like iodoacetamide for chemical modifications which help inhibit re-formation of disulfide bridges and stabilize C residues in the form of the carboxyamidomethyl. After this step, the obtained mixture is filtered through dialysis in ammonium bicarbonate. This removes the alkylation, denaturing and reducing agents, and most likely results in precipitation of proteins. The precipitated proteins are then incubated with trypsin which allows breakdown of all the proteins into their constituent peptides. LC-MS/MS analysis can subsequently be used to carry out protein identification using the peptide mixture.

Table 9.1: Various Chromatographic Techniques used for Isolation of Fungal Proteins and their Respective Advantages and Disadvantages

Chromatographic technique	Advantages	Disadvantages
Ion exchange	High resolution, speed, high protein-capacity resins, crude protein preparations suitable for use	Salt interference, total protein purity rarely achieved
Gel filtration	Separation based on molecular mass, high purity achievable, no salt interference	Low resolution, pre-fractionation required, time-consuming
Affinity	Excellent protein purity achievable, rapid, no pre-fractionation required	Requires affinity ligand, harsh protein release conditions
Immunoaffinity	Excellent protein purity achievable, rapid	Requires protein-specific purified IgG, immunoaffinity-purified protein must be IgG free.

(*Source*: Sean Doyle (2011). Fungal proteomics: John Wiley & Sons, Ltd.)

9.2.5. Recovery of Protein

The fungi produce certain extracellular enzymes within the culture media along with the proteins that are produced intracellularly by mycelia. The variety of enzymes produced depends on the carbon source employed. Fungi can be cultured by both solid-state fermentation and by submerged fermentation. In both cases, the amount of enzymes secreted is usually low, in ranges of ng/mL or mg/mL). Due to this, they have to be concentrated before they analyzed and characterized. Enzymes obtained by SSf are recovered by resuspending them in aqueous buffered solutions (usually containing low concentrations of detergents). Amount of enzyme present in this resuspended material from SSF or the supernatants obtained from submerged culture are influenced by precipitation using ammonium sulfate, ultrafiltration of proteins, lyophilization, and trichloroacetic acid (TCA) precipitation. After concentrated, the volume of the obtained mixture has been reduced by up to 50 to 100 times. Following this, the analysis of protein estimation or protein electrophoresis can be carried on the sample which help estimate and visualize all the proteins present in it. Many fungal species including those belong to basidiomycete species release carbohydrate polymers into their extracellular environments which can interrupt with secretory enzyme isolation. This material is removed using high-speed centrifugation before testing protein concentrations when the secretome (extracted proteins) from fungi are being analyzed.

9.3. ELECTROPHORESIS

9.3.1. SDS-PAGE

One of the most useful and direct technique for obtaining fractionation and separation of proteins is SDS-PAGE. When proteins are incubated with SDS for 5 mins at 95°C, this anionic detergent binds to and denatures them. To this denaturing mixture, addition of reducing agents results in complete denaturation of proteins through disruption of disulfide bridges present at both intra- and inter- molecules. SDS-PAGE works on the principle that when all proteins are fully denatured and reduced are run on a matrix made up of polyacrylamide, they will move towards the positive anode in the presence of an electric field. As result, the migration rate of proteins will depend on their sizes, with smallest molecules being the quickest. Consequently, all the proteins move and migrate depending on their sizes and result in formation of a 'ladder' made up of individual protein within the polyacrylamide gels (Figure 9.1). After electrophoresis is performed on the protein mixture, the proteins need to be fixed and stained on the gel itself. This is done by using staining agents like Coomassie brilliant blue (CBB), silver staining or Amido Black.these stains allow us to visualize protein present in the gel. Luckily, CBB, which is the most commonly employed stain, does not interfere with the subsequent protein analysis on the excised protein bands, such as fractionation by trypsin digestion and MS identification employed on the fractionated proteins. Silver stains can interrupt with the action of trypsin and subsequent LC-MS identification but occasionally, good results can be obtained using this stain.

Western blotting is process used for protein identification after separation of SDS-PAGE. This is done by utilizing antibodies, both polyclonal and monoclonal. The SDS-PAGE gel is obtained which is not stained and is placed sandwiched between membranes made up of either nitrocellulose or polyvinylidene fluoride (PVDF). This arrangement is then exposed to electrotransfer to allow migration of proteins from the gel onto the membrane. The proteins are now replica-transferred onto the membrane, they are then 'blocked' to prevent non-specific binding of antibodies to the membrane and subsequently exposed to immunoreactive proteins which are a mixture of antibody, antibody-enzyme conjugate and substrate in order to visualize them (Figure 9.2). The combinatory use of techniques involving immunology and proteomics for identification of proteins is termed as immunoproteomics. It should be noted that for successful identification of proteins, they are first subjected to SDS-PAGE which yields a gel that is CBB stained (gel-1) and then another gel is obtained which is subjected to western blotting (gel-2). The former technique (yielding gel-1) facilitates mass spectrometry for protein identification and the latter technique (yielding gel-2) allows co-analysis of the protein's immunoreactivity.identification of immunogenic fungal antigens that are linked to several diseases are done using this approach. For example, it is employed for detection of allergy in humans.

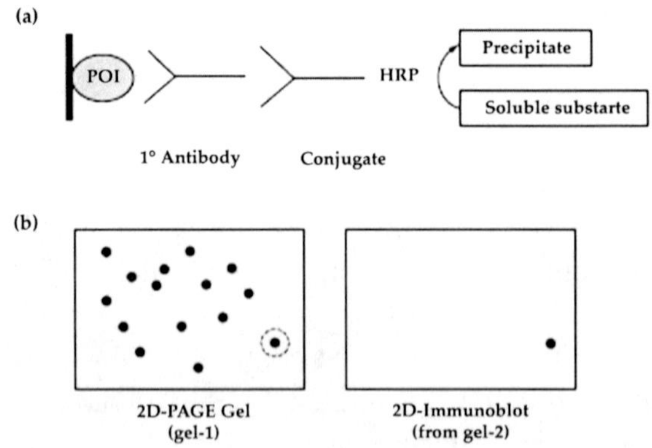

Figure 9.2: Immunoproteomics. A) Recognition of protein under study using purified IgG or a serum antibody. B) replicate 2D-PAGE is carried out and the gel-1 obtained is undergone staining with CBB and electrotransfer is performed on gel-2 onto an appropriate membrane. Subsequently, detection of immunoreactive proteins (2D-immunoblot) is carried out through immunoblot analysis on this membrane by probing it with an antiserum like in (A). The relevant protein found on gel-1 (shown as a circle) can then be isolated through excision and eventually identified using MS analysis. Note: SDS-PAGE can be employed for immunoproteomic investigations alas well even though compared to 2D-PAGE the resolution is low. (Source: *Sean Doyle (2011). Fungal proteomics: John Wiley & Sons, Ltd.*).

In fungal proteomics, SDS-PAGE has proved to be an extremely useful tool regardless of the fact that it is unable to provide a sufficient resolution and separation of proteins present in a mixture of proteins before subsequent MS analysis can be done. One reason is that MS detection is extremely sensitive, such that, it is able to detect proteins that are not visible on the CBB-stained gel. Conclusively, fungal proteins are separated using 2D-PAGE before carrying out MS analysis on them.

9.3.2. 2D-PAGE

This technique basically involves two major steps; IEF and SDS-PAGE. Sometimes it is also termed as 2-DE (two-dimensional electrophoresis). For analysis of fungal proteins, total protein extracts are obtained under native conditions and in the absence of contaminations of salts either from mycelia or from the secretome of fungal cells. These extracts ((250–400 μg) are suspended in an IEF buffer and subsequently placed onto isoelectric focusing strips. They are then subjected to a high-voltage current (around 8000 V) for 24 hours. This results in focusing of proteins as they begin to migrate based on their charges within the electric field unless they reach a point which is equal to their pI (isoelectric point) after which they stop moving. Once this process is completed, the IEF strip is kept beside an SDS-PAGE gel and electrophoresis begins. At this stage, the proteins migrate from IEF strip onto the SDS-gel and then move according to their molecular weights. Once electrophoresis is complete, the protein spots are detected using stains on the gel, such as CBB or the highly-sensitive colloidal Coomassie dye (detection limit: 0.1 μg protein/spot) (Figure 9.3). Sometimes other than these stains, fluorescent dyes, like Sypro Ruby or Deep PurpleTM can also be employed but detection using such dyes requires fluorescent scanners that are able to detect them. Each of these spots consisting of proteins can then be excised through automated or manual methods for further digestion with trypsin and MS analysis. Additionally, fungal proteins can be detected using fluorescent labels such as Cy2, Cy3, and Cy5 before subjecting them to IEF and SDS-PAGE. This method excludes the need for staining of gel after 2D-PAGE and it also makes quantitative analysis feasible according to relative expression of proteins under various conditions of the experiment within a single PAGE gel. This whole process is referred to as 2D-DIGE (2D-difference gel electrophoresis. This process needs the employment of an extensive environment for instruments like fluorescent scanners and automatic protein 'spot-picking' devices. For analysis of fungal proteome, 2D-DIGE has been used for quantitative assessment of relative expression of proteins within the organism. For example, it can be used to identify protective proteins that are expressed under specific environmental conditions like oxidative stress. It can also be used to detect expression of hydrolytic enzymes for detection of various carbon sources.

Similar to SDS-PAGE, 2D-PAGE gels can also be detected by exploiting their immunogenic properties through the use of antisera (Figure 9.2). Immunoproteomic techniques, therefore, can be employed for fractionated proteins obtained after both SDS-PAGE and 2D-PAGE. Interestingly, the results from 2D-PAGE have a higher resolution and sensitivity compared to SDS-PAGE.

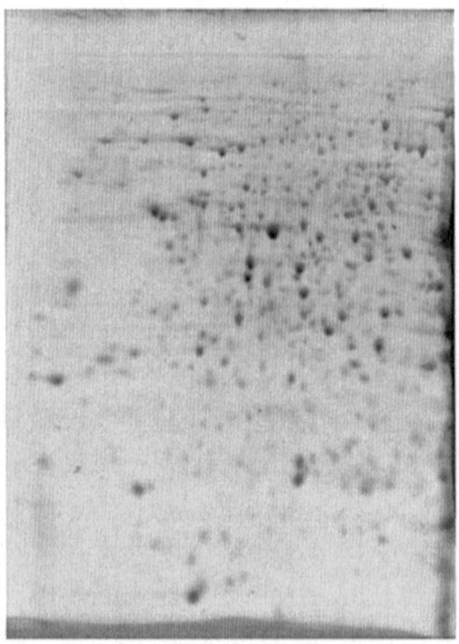

Figure 9.3: A CBB stained 2D-PAGE gel showing protein extract obtained from mycelia of an ascomycete species. To carry out IEF, pH was maintained from 4 to 7 and PAGE was carried out at 10% to allow separation based on their molecular weights. Each spot represents individual proteins from the organisms. The intensities of the spot corresponds to their relative concentrations. (Source: *Sean Doyle (2011). Fungal proteomics: John Wiley & Sons, Ltd.*).

9.3.3. Offgel Electrophoresis

Offgel electrophoresis is a fractionation techniques which is based on solution. It is often used for separation of proteins or peptide before subjecting them to MS analysis. In fact, fractions of peptide that are achieved after this process can be analyzed and identified directly using LC-MS. In fungal proteomic analysis, this technique has not yet been used frequently yet and therefore will not be discussed in detail in this book.

9.4. PROTEIN MASS SPECTROMETRY

9.4.1. Genome Databases

Before we begin discussing MS techniques with reference to fungal proteomics, it will be worth mentioning the vast amount of computational data related to DNA and cDNA sequences from various fungal species that are available online. Researchers worldwide have contributed to these databases by providing and publishing data about nucleic acid sequences present in around 100 fungal genomes. Table 9.2 displays many of these websites' addresses and portals. With the rapid advances in high-throughput next-generation DNA sequencing, there is an increasing amount of data that is being deposited into these locations which are both related to fungal genome and cDNA. This is also due to the realization that fungi are reservoirs for a wide variety of enzymes that are able to degrade carbohydrates and hence can serve for production of biofuels. More significantly, such databases provide in silico data which is raw data regarding DNA sequences which can subsequently be compared to published data related to fungal proteins (e.g., sequences and masses of proteins) in order to facilitate identification of proteins. The success of fungal proteomics and protein identification is majorly dependent on the ability to compare theoretical data about tryptic peptides obtained from putative protein sequences derived from genes present in *in-silico* databases with the mass and sequence data obtained from actual experiments on tryptic peptides obtained from specific proteins (Figure 9.4).

These databases contain data that can be either downloaded to local servers or it can be accessed and interrogated online. Moreover, it is worth to note that even though there is extensive information available for DNA sequences, there is usually very less information provided related to structure on genes, presence of introns and exons, gene splicing, expression or repression of gene, and the post-translational of proteins, if any are expressed. A major issue in data obtained from fungal genomes is that the predicted genes can be classified either as 'hypothetical' or 'predicted' proteins. This is because these proteins are either not identified yet or even if they are detected, no function has assigned to them. The challenge of this 'functional proteomics' is a major problem in fungal biology today.

Table 9.2: Websites having data related to Fungal Genomes Proteomes and General Information

Aspergillus genomes	http://www.cadre-genomes.org.uk/
Multi-fungal genome database	http://mips.gsf.de/genre/proj/fungi/fungal_overview.html
Aspergillus nidulans genome	http://www.broad.mit.edu/annotation/genome/aspergillus _nidulans/Home.html
Aspergillus niger	http://genome.jgi-psf.org/Aspni5/Aspni5.home.html
US Government Genome Sequencing Initiative	http://www.jgi.doe.gov/ http://genome.jgi-psf.org/programs/fungi/index.jsf
Fungal portal	http://fungalgenomes.org/blog/
Fungal portal	http://www.fgsc.net/
SwissProt	http://www.ebi.ac.uk/uniprot/
Expasy	http://expasy.org/
MASCOT	http://www.matrixscience.com/

(Source*: Sean Doyle (2011). Fungal proteomics: John Wiley & Sons, Ltd.*)

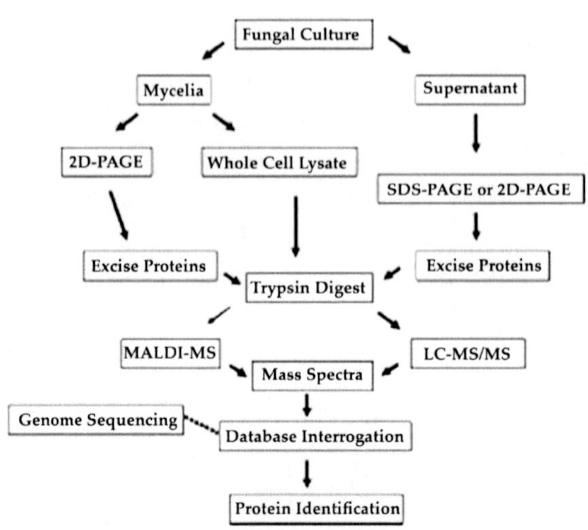

Figure 9.4: Schematic presentation of processes involved in fungal proteomic analysis. From the culture of fungal cells, the mycelia and supernatants are first separated. SDS-PAGE or 2D-PAGe can be employed to allow fractionation of

proteins through electrophoresis or through chromatography (not included in the figure). In another method, the cells can be lysed and their lysates subjected to trypsin digestions without fractionation. Subsequently, information about protein masses and sequences is obtained by using the mixtures of peptides which have been derived from either purified individual proteins or from mixtures of proteins (through shotgun proteomics). This is achieved by using LC-MS/MS or MALDI-TOF MS. The data obtained in then interrogated against cDNA or genomic databases that are publicly available to allow identification of proteins. (Source: *Sean Doyle (2011). Fungal proteomics: John Wiley & Sons, Ltd.*).

9.4.2. Fragmentation of Proteins

Before a protein or protein mixture purified from gel is subjected to MS analysis, it has to undergo digestion using trypsin in order to obtain its constituent peptides. Trypsin is used for this purpose because it allows fragmentation of proteins at the C-terminus of arginine or lysine and produce a peptide population, each ending with a positively charged amino acid. This positive charges facilitate the flight of peptides in matrix-assisted laser desorption ionization–time-of-flight mass spectrometry (MALDI-ToF MS or MALDI MS). In addition, tryptic digestion yields peptides that have masses in the ranges between 500 to 2500 Da, this is feasible range for optimized detection in instruments used for MS analysis. Trypsin that is employed is always high-grade, which means that it is free from protease contaminants and furthermore, it is treated with certain chemicals that inhibit auto0digestion by inhibition of endogenous chymotryptic activities. Subsequent detection of peptide by MS is interrupted and the efficiencies reduced if any of these activities are occurring. In this section, we will not discuss details of the procedure used for tryptic digestion. However, it is worth mentioning that the target proteins can be obtained in the ranges of nanogram/microgram to milligram. The target protein_trypsin ratio is usually 20:1. Theoretically, subjecting proteins to tryptic digestion should results in generation of all the tryptic peptides. In practice, this is rarely achieved. From the perspective of its chemistry, this inefficient detection of all constituent tryptic peptides may occur due to the following reasons:

1. Cleavage of peptide bonds in inhibited;
2. Large peptides that are hydrophobic tend to adhere to plastic or are insoluble resulting in their loss;
3. Peptides that are too short, having molecular masses lower than 300 Da are hard to detect;

4. Prior to MS analysis, the peptides are exposed to filter to allow preparation of samples which results in non-specific binding to filters;
5. Residues of lysine and arginine that are modified;

From the perspective of the instruments used, some of the sample peptides are not ionized in an efficient manner during MALDI-ToF MS analysis. Some of them can also bind irreversibly to LC columns that are used for LC-MS, hence cannot be detected;

9.4.3. Mass Spectrometry

It is to be noted that when protein MS is referred, it actually involves protein identification by comparing the obtained peptide data with cDNA or genomic data present in databases. Proteins are therefore identified depending on the accuracy of the data related to masses and sequences of the peptides and determination using MS analysis. Accurate determination of peptide or peptide fragments' masses can be achieved using modern mass spectrometers. Subsequently, this data is compared against the data present from theoretical tryptic peptide masses or their corresponding sequences obtained from putative protein sequences derived from cDNA or gene sequences present in *in-silico* databases (Figure 9.4). In mass spectrometer, peptide detection is possible only when they are ionized by employing either laser or electrospray methods.

9.4.4. MALDI-ToF MS

This is a specific type of protein MS used for identification of proteins. In this technique, a pure form of protein is digested to release peptides which are then mixed with a material, such as α-cyano-4-hydroxycinnamic acid (HCCA). They are then targeted into a metal plate and allowed to dry. Once dried, the sample is subjected to ionization using laser light within the vacuum chamber of the mass spectrometer under high voltage. Energy from the laser is transferred to the constituent peptides with the help of the matrix (HCCA). As a result, the samples form vapors that have a positive charge on them. Due to electrostatic forced within the time-of-flight detector, these molecules are migrated according to their mass: charge ratios also referred to as m/z ratio. This yields a fingerprint of peptide masses or a MS spectrum as shown in Figure 9.5). It is not possible to detect all the constituent peptides in the proteins using this method and it has the ability to detect intact proteins and other biomolecules such as carbohydrates and nucleic acids. For such analyses, the matrix material that needs to be used and the strategies used for depositing targets will vary accordingly. One other example of matrix material is 2,'5'-dihydroxybenzoic acid or sinapinic acid. MALDI-ToF MS technique offers several advantages like possibility to analyze several specimens simultaneously or sequentially in as less time as 5 mins, thereby providing a very quick an automated approach. However, this technique is not as sensitive as other MS techniques used and the protein sample that can be identified with sureness needs to be relatively pure.

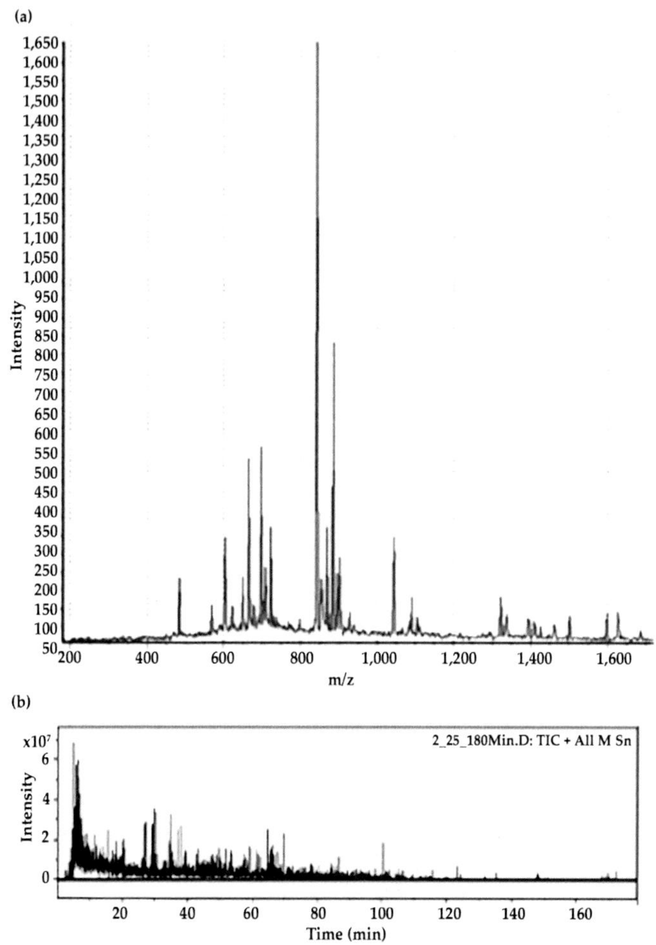

Figure 9.5: MALDI-ToF MS spectrum and LC-MS/MS total-ion chromatogram. (a) The MALDI-ToF MS spectrum is a plot of *m/z* value of the positively charged peptide ions against signal intensity, and optimized results are obtained in the *m/z* range 500–2500. The *m/z* values represent the precise mass of the

peptide ions derived from the protein of interest and are automatically used to compare with the tryptic peptide data published in the *silico* databases predicted using sequences from DNA and cDNA. Protein matches are usually displayed as (i) percentage of sequence coverage or (ii) number of peptides observed/ number of peptides preditect. The greater the score, higher will be the accuracy of the identification. B) Shotgun proteomics. In the ions chromatogram shown the ion intensity is plotted again time in mins. It represents all the peptides that are produced through digestion by total fungal proteins lysate and were subsequently detected using LC-MS MS. Optimal fractionation of peptides is achieved when 1.5 mg of sample is used and the runtime allowed for LC is 180 mins. Software employed in instrument and the expertise of the operator help to correlated the *in silico* sequences of DNA and cDNA that are translated with the data about mass/sequence of peptide which enables successful identification of proteins. This technique helps in identifying hundreds of proteins present in a single sample. (Source: *Sean Doyle (2011). Fungal proteomics: John Wiley & Sons, Ltd.*).

9.4.5. Electrospray Mass Spectrometry

In contrast to MLDI-ToF MS, electrospray or nanoelectrospray techniques produce molecular species like peptides from liquid state directly to those that are in gaseous state and are ionized. This technique is also referred to as nanoelectrospray ionization (ESI or nanoESI). In this technique, the separation of peptides using LC can be directly linked to analyzers of high-mass accuracy through ESI. This has resulted in their increasing applications during the past decade. The sample of peptide mixture that may be pure or relatively pure or it may be a mixture of peptides obtained from protein digest of complex mixtures is first fractionated using phase high-performance liquid chromatography (HPLC). At the end of LC column, the peptides are eluted out and subjected to ionization using nanoESI or ESI. Subsequently, the masses of these ionized molecular species are accurately determined before protein identification through comparison with the data available *in silico* databases. In this LC-MS approach, the proteins are identified very accurately using a very small quantity of starting material, therefore making it preferable over MALDI-ToF MS. In addition, the protein sample is purified during LC and therefore the additional purifying step that is employed in MALDI is skipped. Finally, the duration for which LC-MS needs to be optimally just 30 mins which is lesser compared to MALDI-ToF. An improved identification for the sequence of peptide is attained by comparing the pattern of theoretical peptide fragments present in the in silico databases with pattern of sub-fragmentation for peptides observed in actual experiments. The protein hits observed are therefore with higher specificity and confidence. Such analyses are not carried out manually but it is performed industrially using software which are available as standards with other commercial instrumentation.

9.4.6. Shotgun Proteomics

Until now we have focused on techniques that require digestion of pure or partly pure proteins using trypsin before they are analyzed using mass spectrometry. However, recently there have been improvements in fractionation and peptide separation using LC and more necessary, enhancement of algorithms employed for assignment of detected proteins to a corresponding database protein has resulted inability for simultaneous identification of proteins constituents found in complex mixtures through MS analysis. The term used for this kind of approach is shotgun proteomics. It eliminates the need for trypsin digestion and subsequent analysis using MS once the proteins are purified. The method used for analyses using shotgun proteomics is described in Figure9.4. The mixture obtained after trypsinization consists of peptide fragments that are produced from all the proteins that are found in the initial sample and thus, it yields thousands of different peptides. Subsequent fractionation followed by LC-MS/MS allows generation of proteins that have been identified by comparing with published data in relevant *in silico* databases. Identification of some proteins is possible using data that is as simple as a single peptide, however, majority of the cases require analysis of two peptides from each protein, or 5–10% coverage of sequence, in order to allow constituent protein identification confidently. Even though a large quantity of data is yielded from each experiment in shotgun proteomics, it can take a lot of time to carry out each run in LC (2 to 3 h) and to analyze sequence data (Figure 9.5).

9.4.7. Quantitative Proteomics

The above-mentioned techniques used for fungal proteomics basically provide qualitative data. Meaning that the protein presence is detected but there is no estimation for its quantity present under certain conditions and how this quantity is influenced due to alterations in various conditions. Usually, quantitative proteomics involves using analysis from 2D-PAGE obtained from experiments carried out in two different conditions and is followed by detection of differential protein expression through image analysis. Nevertheless, strategies are available which are gel-free. Such methods are done using proteins that are chemically labeled on cysteine (isotope-coded affinity tag (ICAT)) or lysine ((isobaric tag for relative and absolute quantification (iTRAQ)). This is followed by trypsinization and protein analysis using LC-MS/MS. Proteins can be quantified through a comparative mass spectrum using computerized intensities of ion signals on pairs of peptides that are labeled differently. This technique is particularly useful tool for analysis of fungal proteomes as it helps in following ways:

(i) Investigation of effect on expression of proteins due altered culture conditions;

(ii) Studying the effects introduced due to deletion of genes on both specific and global expression of proteins; and

(iii) Determining changes in expression of virulence factors as a result of infections.

9.5. PROTEOMICS OF FUNGI

9.5.1. Trichoderma Proteomics

The mycoparasitic fungi, Trichoderma harzianum, provide protection to plants from pathogenic fungal species such as Rhizoctonia Solani and Botrytis cinerea. It, therefore, functions as a biocontrol agent. In addition, these species secrete proteins that belong to a wide spectrum, especially those that are degradative enzymes for carbohydrates. The community of fungal proteomics has therefore been highly interested in the Trichoderma spp. Before 2004, there was not much data related to Trichoderma harzianum proteome. Initially, investigations related to its proteome were carried out using fractionation by 2D-PAGE and subsequent protein identification using LC-MS/MS and MALDI for 25 proteins that were intracellular, such a glycolytic enzymes and heat-shock proteins. After this data was obtained, the work was extended to analysis of proteins that were expressed when exposed to extracts of cell wall of the pathogenic fungi, *R. solani*. This examination was done using Trichoderma atroviride (formerly *T.harzianum* P1). Through this approach, it was found that there were eight enzymes that were produced in *T. atroviride* that were highly expressed. These were detected using MS analysis and the proteins identified were N-acetyl-β-D-glucosaminidase, endochitinase, vacuolar protease A, superoxide dismutase, trypsin-like protease, a serine protease and a hypothetical protein. Immediately it was observed that most of these proteins belonged to categories of enzymes that were secreted either for hydrolysis or for protection and were therefore secreted most probably for degradation of biomolecules present in *R. solani*, such as, material constituting the cells wall including its proteins. This implied that extracts from fungal cell walls of a particular species have the ability to induce a biological response in another fungal species. Such responses and interactions can be studied using proteomics. Interestingly, the enzymes produced by *T. atroviride* were located at different spots on the 2D-gels. These multiple spots correspond to the various charged forms of an individual protein. This can be especially obtained using 2D-PAGE because of the high resolution obtained through IEF thus allowing this separation. Interestingly, this technique also detected proteins that were hypothetical. This means that the protein did not have any known function or homology to an already known protein. Hypothetical proteins are often detected in fungal proteomics and therefore, as explained earlier, pose a challenge in future of fungal proteomic and genomic studies. Proteomics can also be used as a tool for analyzing differential expression of proteins in mixtures of cultures containing cells from *T. atroviride*, *B. cinerea*, and plant leaves. Such mixed-proteomic analysis exploits the MS technique's high specificity of detection to allow identification of proteins that are generated by each of the three types of cultures under the conditions which mimic biocontrol and infection conditions. For example, when *B. cinerea* is co-incubated with leaves from bean plant and *T. atroviride,* it results in up-regulation of cyclophilin A (a protein functioning in folding of proteins) and superoxide dismutase (a protein that acts as scavenger of free radicals).

Prior to proteomic studies, the cells or cell lysates can be used for enrichment of subcellular organelles (such as nuclei or mitochondria) or specific fractions of proteins (such as proteins with high molecular masses). Sub-proteomics is the term used to describe this strategy for pre-enrichment. In *Trichoderma* species, such sub-proteomic studies have been done using protein MS. Specifically in *T. harzianum*, the proteome present in mitochondria and an is 26S proteasome found in commercially significant *Trichoderma reesei* species was elucidated through electrophoresis and MS analysis. These findings were based on the fact that before fractionation and identification of proteins, the organelles were purified. This facilitated the detections of proteins that were expressed less abundantly and were highly localized within sub-cellular compartments in the fungal cell. Several proteins were detected through this study, which were found to have functions in regulations, rather than catalysis, unlike *Trichoderma* spp. Because *T. reesei* is a source of numerous commercial enzymes it is necessary to study its 26S proteasome, the degradative system in this microorganism. This will confirm the circumvention of factors that inhibit stability of proteins and their release. This will allow optimal production of proteins in order to enhance the amounts of secreted enzymes.

9.5.2. Aspergillus Proteomics

One of the most significant human pathogenic fungal strain is that of *Aspergillus fumigatus*. It harms patients that are specifically immunocompromised. Other Aspergillus species including *Aspergillus niger* and *Aspergillus oryzae* produce important enzymes and metabolites respectively that are used in food industry. Several genomes belonging to seven of the species belonging *Aspergillus* genus have been sequenced and are available online. Their sizes are in the range from 28.6 to 36.7 Mb. Because there are many genes that have been identified from these data that are not associated with any function yet, it is necessary to apply both genomic and functional studies for these proteins in order to find out their biological roles. Investigation of *Aspergillus* spp. is difficult and proteomic studies illustrated just 10 published data related to 2D-PAGE and MS analysis during the years from 2002 and 2007? There is an increasing popularity of these significant species in biomedical and industrial applications. The researches for using these species for diagnostic purposes, novel targets for drugs and isolation of important enzymes have intensified recently and as a result, there is an increasing number of publications emerging recently.

The intracellular proteome of *A. fumigatus* has been studied by employing proteomic and immunoproteomic studies. It has help elucidate secretory proteins that are produced by the species when exposed to different culture conditions. Moreover, several antigenic, immunoreactive, conidial- and biofilm- linked proteins synthesized by the organism have been found. For *A. fumigatus,* methods have been developed for optimized extracting of proteins and for investigating its intracellular proteome and has been the target of many researches. Eventually, many groups of scientists have found hundreds of intracellular proteins produced *in A. fumigatus* and among this proteome, there are many which have unknown functions. These proteins predominantly fall into the category of glycolytic enzymes, catalases, heat-shock proteins and mitochondrial enzymes. In one such study, cell lysates from *A. fumigatus* were first pre-enriched and their sub-proteomes analyzed using affinity chromatography. This revealed the present of activity of glutathione *S*-transferase which is a putative protein acting as an elongation factor for translation. In another instance, prior to 2D-PAGE and MALDI

ToF MS, enrichment was done through gel permeation chromatography. This resulted in detection of a high-molecular-weight siderophore synthetase which was produced in low amounts as a result of iron depletion condition.

Many researchers have worked to identify the mechanisms involved in *A. fumigatus for producing a tolerance or resistance response to drugs used against fungi. Responses due to exposure to* voriconazole, caspofungin or amphotericin B have been investigated in terms of fungal proteome. The key to outstanding rather understanding the mechanism for drug resistance in fungi is through finding altered expression of proteins that are both intracellular and extracellular. Moreover, these findings help the scientists to enhance the potential of these antifungal drugs by developing strategies for overcoming resistance elicited by these proteins. For example, exposure to amphotericin B to fungi resulted in altered expression of around 85 proteins (76 were upregulated and 9 were downregulated) by at least two-folds relative to normal growth conditions. Among these proteins, cell-stress proteins, enzymes, and transport proteins were included that were involved in the biosynthesis of ergosterol, which is a target of amphotericin B. In addition, changes were also detected in enzymes that were involved in secretion of proteins and it is yet unclear how significant this finding is.

Biofuel production (e.g., ethanol) using lignocellulose biomass is also possible by the employment of hydrolytic enzymes and fermentation. However, it is not prevalently used due to its expensiveness and relatively low efficiency of the employed enzymes that function to breakdown cellulose into fermentable sugars. The proteome of *Aspergillus* spp. has been analyzed to identify enzymes that have potential to degrade carbohydrates in order to generate substrates for production of biofuel. In one such study, *Aspergillus fumigatus* was found to produce a thermostable β-glucosidase enzyme that is secreted and allows degradation of cellulose in a more effective way relative to other *Aspergillus* spp. enzymes. Typically, fungal strains consist of many enzymes that are commercially significant and have used in several processes in biotechnology as discussed earlier in this book.

In conidia of *A. fumigatus, 57 proteins have been identified through 2D-PAGE and MALDI-ToF MS that were overexpressed compared to those present in mycelia of fungi. Moreover, enzymes involved in anaerobic fermentation, such as* pyruvate decarboxylase and alcohol dehydrogenase, were found to be expressed which implies that in resting condia, there is anaerobic metabolism. In addition, enzymes associated with resistance to stress were also produced. These include those involved in biosynthesis of pigments, in synthesis of surface of conidia, and to inactivate reactive oxygen species (ROS). It is predicted that they help in protecting the condia of *A. fumigatus* through innate and adaptive immune systems. The stages of germination of conidia in *Aspergillus nidulans* has also been studies through employment of proteomics. It has been revealed that 241 proteins were differentially expressed ($P < 0.05$) during the early stages of germination. Among these, 40 out of 57 proteins detected by MALDI-ToF MS were found to be linked with detoxification of reactive oxygen species, metabolism of energy, synthesis of proteins, and folding process of proteins. The altered expression of proteins in the conidia of *Aspergillus nidulans* were identified simultaneously and confirmed through analysis using northern blots and reverse transcriptase (RT)-PCR. However, it is worth noting that correlation between levels of genes and proteins are not observed every time when genome and proteome of fungi are analyzed.

When genes are identified using high-throughput bioinformatics tools, the accuracy of the predicted genes is confirmed through proteomics. This is referred to as gene or gene model calling. This is due to the fact that detection of tryptic peptides occurs as a result of transcription and translation of actual genes. In fungal genes, there is presence of multiple introns which makes it difficult to identify the start and stop codons present their sequences even when very efficient bioinformatics tools are used. Several reliable tools *in silico*n help in the prediction of splice sites (found in the intron:exon regions) with confidence level of up to 95%. However, it is not always possible that the cDNA sequences of fungal genes are predicted correctly in silico. Because a protein sequence corresponds to the cDNA sequence, the resultant tryptic peptides released will, therefore, be derived from a sequence of DNA that contains exonic, intronic, and exon-intron splice regions (Figure 9.6). Sometimes, sequences of tryptic peptides contain regions from internal exons and hence, they must be generated from fusion of exon:exon. This, therefore, confirms that the splice sites are existing between the exons and introns and implies that the prediction software for splice sites are accurate (Figure 9.6). These findings have been concluded with the help of LC-MS/MS analysis employed to tryptic peptides which are generated from proteins belonging to *A. niger* that are separated by using SDS-PAGE only. In this investigation, mapping of tryptic peptides was done against genome sequences from two *A. niger*strains. One *of them was ATCC1015 which was sequenced by* the Joint Genomes Initiative (Table 9.2). CBS 513.88 was the other one which was sequenced by the Dutch biotechnology company DSM. A total of 405 sequences of peptides were yielded from 19628 mass spectra. These were mapped against 214 locations of the *A. niger* genome. Among this, 6 % were not present in the predicted model of the gene. Resultantly, the generated peptide data was employed for modification or correction of in silico connotation processed of the genome and it also helped in confirmation of around 54 predicted sites for intron-exon splicing. This approach involving experiments demonstrates the use of proteomic data invalidation of annotation of fungal genome, identification of intron: exon splice sites and translation of mRNA.

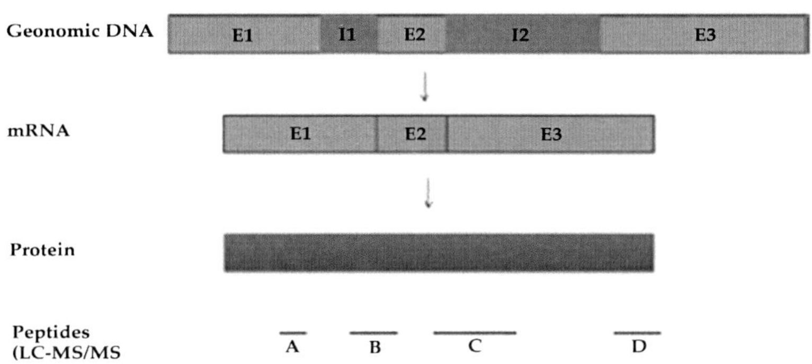

Figure 9.6: Confirmation of *in silico* predicted slice sites present in fungal genome through protein MS analysis. The gene of interest was analyzed by using fungal genome sequencing and it was found to containing two introns (11 and 12) and one stop codon within the Exon 3 (E3). Tryptic digestion followed by LC-MS/MS analysis of the protein showed the presence of four peptides (A-D). Peptide A was originally present within exon 1. Peptides B and C were detected only due to presence of intron 11 and 12 within the DNA and their subsequent excision during transcription of the respective mRNA sequences. In peptide D, there were no K or R residues present in the C-terminus and therefore detection through LC-MS/MS analysis strongly implies that the correct stop codon (*) was annotated within the genomic DNA. (Source: *Sean Doyle (2011). Fungal proteomics: John Wiley & Sons, Ltd.*).

When *A.niger* was cultured on different sources of carbon, the secretome synthesized was explored through fractionation of proteins using SDS-PAGE, followed by LC-MS/MS shotgun proteomics. These were further interrogated against fungal genomic databases. As a result of this study, 200 secretory proteins were identified that were encoded by signal peptides that predicted *in silico*. Changes in secretomes of this species were also revealed due to alteration in culture conditions. In this case, secretion of proteases was increased when there was a lack of carbon sources. Moreover, it was found that secretion of pectinolytic enzymes was up-regulated as a result of using galacturonic acids as the carbon source. Interestingly, there was detection of a protein within the secretome which had similarity to Asphemolysproduced in *A. fumigatus* and which did not contain a signal peptide.

A. niger has several uses in biotechnology at industrial level. This is due to the fact that it is highly potent in secretion of proteins through a process which involves intracellular organelles that function especially for secretion. This helps greatly in downstream processing and in recovery of enzymes. In fact, the proteins secreted by this filamentous fungus have been analyzed using both transcriptomic and proteomic analysis in order to study factors that affect their secretion. In this section of the book, however, we will be focusing only on the proteomic studies. In one investigation, the secretion of cellulase and hemicellulase which is induced in the presence of D-xylose was studied. The composition of proteins present in secretory organelles of *A. niger* were investigated and compared against those present in cultures containing sorbitol for induction. Subcellular fractions were isolated which contained microsomal membrane fractions that were rich in Golgi and ER components. This was achieved by using shotgun proteomics that employed LC-MS/MS. Through this approach, more than 1000 proteins were identified which were thought to participate in secretion of proteins. Among these proteins, the most abundant ones were chaperones, disulfide isomerases, mannosyltransferases, and GTPases. Interestingly, when D-xylose was used in culture for protein secretion induction, then the 20S proteasome complex was recruited towards the microsomal fractions. This suggested that this complex was involved in induction of proteins under specific conditions. For such analysis, quantitative PCR and microarray techniques would not have been helpful tools because the expression of genes coding for components of 20S proteasome was not altered in this case. Therefore, this interaction could be revealed only through this approach. Conclusively, this study demonstrated the function of 20S proteasomal complex in secretion of proteins and it also implied that *in vivo* protein-protein interactions can essentially be elucidated through proteomics.

Through LC-MS/MS analysis, several proteins with unknown functions, referred to as hypothetical proteins were detected. Proteins with unknown functions were detected and this confirms that cognate genes were expressed and it also helps us hypothesize that these proteins may be significant in the process of protein secretions. Such hypotheses can be tested through investigations involving gene knock-out techniques in which the gene of interest can be specifically deleted yielding mutant phenotypes. In this case, mutant phenotypes related to secretion of proteins will be detected and assessed relate to *A. niger* that is wild-type. If there is any change detected in protein secretion in the mutant strain, the gene function can be assigned a role. The secreted proteome of wild-type *A. niger* has also been analyzed quantitatively using iTRAQ along with LC-MS/MS. Even though all the details of the used mutant *A. niger* strain were not available, the increase in secretion of several proteins relative to wild-type strains was detected which were statistically significant ($P < 0.001$). Among these proteins, there were cellulases, hemicellulases, glycosidases, and pectinases. This technique illustrates the high potential of gene knock out techniques for positive influence on global secretion of proteins, when used along with quantitative proteomic analyses that are highly sensitive. Together, they help us to evaluate the influence of loss of gene in secretion of proteins.

9.6. SPECIALIZED PROTEOMIC APPLICATION IN FUNGAL RESEARCH

In addition to the above-mentioned applications, proteomics plays a role in other aspects of fungal research as well. These include identification of post-translational modifications (PTMs), whole-protein or 'top-down' proteomics, and intact conidial proteomics which are worth discussing in this context.

In fungal genomes, there are many types of post-translational modifications present including but limited to phosphorylation, phosphopantetheinylation, glycosylation, ubiquitination, methylation, and acetylation. Analysis of DNA sequences can help us predict the sites for PTMs or motifs that are found within the proteins. However, it is necessary to carry out experiments for biochemical analysis of proteins in order to confirm the presence of the predicted post-translational modifications in vivo. Within fungal cells, ubiquitination is significant in directing protein degradation, glycosylation is important for providing stability to proteins and to localize it to targets, and phosphorylation is usually required for regulation of activation of protein-protein interactions. In scientific literature, their methods employed for PTMs detection are extensively found.

In fungal proteome, acetylation and phosphopantetheinylation are specifically found to be significant PTMS. The latter is the key to biosynthesis (through non-ribosomal protein synthesis) of several natural products and secondary metabolites, such as siderophores, antibiotics, and toxins. The former is important for regulation of activation of chromatin and therefore for fungal gene clusters expression. Acetylation can easily be detected within the peptides through mass spectrometry analysis of proteins due to the detectable mass of acetyl group (42 Da). 4'-phosphopantetheinyl transferase (4'-PPTase) help is activation of Non-ribosomal peptide synthetases by a post-translational modification that attaches to its specific S residues, 4'-phosphopantetheine (present in the coenzyme A). MS also has the potential to detect the peptides that are derived from Non-ribosomal peptide-synthetases and therefore consist of 358.09 Da 4'-phosphopantetheine on them. In A. fumigatus proteomic studies, this approach has been used successfully for confirmation of activation of synthetases and detection of 4'-PPTase activity.

MS analysis can also help detect PTMs present on whole protein samples. Fourier-transform ion cyclotron resonance (FT-ICR) MS is the name given to the technique used for studied proteins that are intact. This is because it allows detection of very minute changes in masses which occur due to PTMs before trypsin peptides are generated. The prevalence of histone acetylation and deacetylation within the fungal proteomes have been determined using FT-ICR MS. Moreover, it has also been used for identifying the types of amino acids that are linked to phosphopantetheine arms of non-ribosomal peptide synthetases, thereby aiding in formation of peptide through non-ribosomal methods.

MALDI-ToF MS analysis has also been used to produce mass spectra from proteome present in intact conidia. This helps in strain typing which is dependent on the pattern generated which is specific to species or strains. In this approach, MALDI-compatible matrices are mixed with fungal conidia and subsequently spotted onto target plates to induce ionization. The matrix materials which allow ionization of the sample used in this technique, called as MALDI-ToF intact cell mass spectrometry (MALDI-ToF ICMS) are 2,'5'-dihydroxybenzoic acid and sinapinic acid. The generated spectrum allows identification of species such as *Penicillium, Fusarium*, *Aspergillus, or Trichoderma* because it is unique for species. Moreover, it can also reveal the strain type as it may be specific to particular strains as well like *Fusarium* spp. In MALDI-ToF ICMS, several ions which are specific to species and fall within the range of 1500 to 15000 *m/z* values are detected, but this does not facilitate identification of actual compounds and their nature which are currently unknown.

9.7. CONCLUSION

Fungal proteomics has been used for a long time now. As the technologies used in proteomics have become advanced, more fungal proteins produced at both individual and large-scale have been identified. This is also due to simultaneous progress in availability of more and more fungal genomic sequences. Together, these have contributed to provide ideal conditions for studying fungal proteomes. It has clearly been demonstrated that proteomes for fungi are very dynamic and illustrate high responsiveness to their external environment. These changes can be explored and studied using tools involved in proteomics. Therefore, research in fungal proteomics is expected to improve our understanding about both the biology of fungi and the mechanisms helping is causing diseases and to help us provide insights into novel enzymes that may be exploited in biotechnology at commercial levels.

10
CHAPTER

APPLICATIONS OF FUNGAL BIOTECHNOLOGY IN AGRICULTURE

CONTENTS

10.1. INTRODUCTION

To maintain our ecosystem and a sustainable agricultural system, it is necessary to avoid the use of fertilizers and fungicides. Recently, there have been a lot of researches in the field of development of bio-fertilizers and biocontrol systems which pose minimal harm to our environment. Naturally, some fungal species are known to exist in mutualism with plant species and are known to have a positive influence on plant growth and performance. These microorganisms are mainly found to be associated with the roots of the plants and are referred as plant growth-promoting rhizobacteria (PGPR) or plant growth-promoting fungi (PGPF). They protect plants by suppressing diseases. In addition, mycorrhizal fungi and endophytes also elicit similar responses in plants with the latter inducing growth in plants. They are most likely to suppress diseases in host plants by using the following mechanisms;

1. Antagonistic activities: They may show hyperparasitism, antibiosis, and competition. This may be done by production of metabolites that enhance antibiotic activity. In some cases, there is competition for colonization sites, direct antibiosis, nutritional aspects, and plant defense reaction;
2. Induction of systemic resistance: This is evident by activation of plant cell defense responses in some plants through lignin deposition, oxidative species burst, accumulation of phenolic, callose deposition, thickening of walls that contain nonesterified pectins and PR-1a protein. Moreover, some enzyme activities are also increased in response to fungi, for examples chitinase andβ-1,3-glucanase.

10.1.1. PGPF

Some fungal strains isolated from soil are known to inhibit plant growth and are referred to as plant-pathogenic fungi, for example, *Penicillium janthinellim* and *P. citreonigrum*. However, there are some fungal species like *Trichoderma sp., Rhizoctonia solani* that are shown to promote growth in plants. Other than the soil-biome these PGPF may be found in other ecosystems as well. The PGPF that belong to *Rhizocantonia*, *Trichodermal*, and some mycelial fungal species have been studied extensively and have shown to have promontory effects on growth of various plant species. These species achieve this through multiple mechanisms. For instance, when treated with PGPF strains of *Trichoderma*, plants were shown to promote growth through induction of gibberellin production in the plants under study. In other cases, PGPF strains are able to increase the amounts of minerals such as nitrates and ammonium ions in plant species such as barley suggesting that mineralization of organic substances by PGPF is associated with their growth promotion effects. Another mechanism that PGPFs exhibit is by suppression of plant-pathogens such as *Phythiumspp* in cucumber. In one case, PGPG help in plant growth by suppression of soil-borne diseases caused by P. ultimatum in cucumber, tomato, barley, and radish plants. When PGPG are applied to roots and leaves of plants through systemic resistance, suppress air-borne pathogenic diseases caused by Colletotrichum orbiculare, for example. Therefore, these organisms can also induce resistance in host plants against diseases and promote plant growth.

10.1.2. Mycorrhizae

Myccorhizae have been defined as "a symbiosis in which an external mycelium of a fungus supplies soil-derived nutrients to a plant root." Some plants are highly dependent on myccorhizae for their survival. These fungal species can support plant life by helping them survive in regeneration niches, enhance production of seeds, quality of seeds, and host and offspring vigor, improve the resistance of plants against environmental stress. In this section, we will discuss the most prominent systems of mycorrhizae, Ectomycorrhizae, and Arbuscular Mycorrhizae

Ectomycorrhizae includes more than 5000 fungal species that belong to Asco and Basidiomycete classes including Laccaria laccota, Pisolithus tinctorius, Suillus plorans, Hebeloma cylindrosporum, and H. crustuliniforme. They are found to be linked to more than 2000 woody plants species such as Douglas fir, prine, and eucalyptus. These fungal species help plants take up water and nutrients and therefore promote growth. Studies on several host plants supported the hypothesis that there was an increased nutrient uptake (such as P, N, K, and Ca) due to inoculation with these mycorrhizal species. The Ectomycorrhizal fungi mostly achieve this through hyphae development which increases the plant's absorptive and assimilation capacity.

Arbuscular mycorrhizal (AM) species include those belonging to Zygomycetes group of fungi. Several fungal species have been shown to promote growth of a wide variety of plants such as tomatoes, gerbera, maize, and asparagus. They do so by enhancing the potential of roots to absorb water and nutrients (such as phosphorus) and by enhancing structure of soil to allow aeration and penetration of water. In some cases, AM fungi have an influence on root morphology and more prominently on host plant's physiology. It causes increases in amount of growth-regulating compounds (such as cytokinin, auxin, and GA), rate of photosynthesis, and alterations in partitioning of photosynthate to shoots and roots.

10.1.3. Endophyte

An endophyte is defined as; "organisms that symptomlessly colonize the living, internal tissues of their host, even though the endophyte may, after an incubation or latency period, cause disease." These may include fungi that are pathogenic to fungi, like smut fungi. Generally, endophytes induce resistance in host plants to damages by insects, environmental stress, and diseases by nematodes and other pathogens. In one study, plants infected with Fusarium moniliforme were shown to increase in weight, shoot height and diameter due to this fungus presence. In another case, *Phialocephala Fortini* improved phosphorus and nitrogen uptake in *Pinus contorta.* Like PGPF and myccorhizae, endophyte also act as biocontrol agents by production of defense compounds during infection of host plants. These belong to *Fusarium* species and protect plants like celery, watermelon, cucumber, and other important crop plants. One particular example is that of *Heteroconium chaetosprira* which suppresses infection by *Plasmodiophora brassicae* on Chinese cabbage.

10.2. MYCOHERBICIDES

Bioherbicides are described as the intentional exploitation of plant pathogens that are mass-produced, formulated, and applied in a similar way as chemicals. A wide range of microorganisms are included in this class including fungal pathogens, which are termed specifically as mycoherbicides. They are used to control the growth of weeds below a threshold level so that they do not interfere with the growth of crops at industrial level. There are several social, environmental, and economical benefits linked to biological control, however, there are also many limitations to their use with aspect of biological, technological, and environmental constraints.

In the last two decades, around eight bioherbicides have been made commercially available and many are under various stages of development. For development of bioherbicide, it is desirable that it exhibits these traits; (1) a narrow host range, (2) ease of use, (3) genetic stability, (4) ability to mass produce inoculum cost-effectively with long shelf life, and (5) ability to be fast-acting with predictable field performance and provide sufficient weed control comparable to chemical herbicides.

10.2.1. Challenges in Biological Control

It is big challenge to find out candidates for bioherbicides that show all of the above-mentioned characteristics. There are several factors that contribute to limit the prevalent development of bioherbicides which will be discussed in this section.

10.2.1.1. Biological Factors

The weeds that commonly grow in crop plants have a huge genetic diversity. Due to this wide-spread variation, it is difficult to develop a herbicide that acts uniformly on all species of weeds. Moreover, having highly specific bioherbicide ensures that it doesn't act non-specifically on important and valuable plants growing around the weeds. Another factor is the architecture and morphology of plants. For example, grass weeds are difficult to be controlled because their meristem in protected from pathogen attack by a sheath. Another example is that on perennial weeds which propagate in rhizomes and stolong making longer-term control of weeds difficult. Other physical barriers are waxy cuticles, hair on leaves which inhibit the action of bioherbicides.

10.2.1.2. Environmental Factors

Temperature and moisture are two main environmental factors that affect the potency of mycoherbicides. A combinatory effect of both factors is often observed rather than each one alone. Germination of fungal pathogens depends on amount and duration of moisture and wetness on leaf. It also affects its penetration structures and therefore, infection on plant. Mycoherbicides can usually function in a wide range of temperatures but the amount of wetness greatly affects it potential. Intermittent wetness of leave for long time periods is detrimental for fungal germination. Some fungal species like rust fungi are dispersed by wind and therefore need less moisture compared to water0dissemianted pathogens that release spores within a mucilaginous matrix.

10.2.1.3. Technological Factors

Production of bioherbicides is commercially possible only if it can be produced stably, viably and on large-scale. For this to be achieved, submerged liquid fermentation is employed for mass-production of biocontrol agents. methods have been developed for specific bioherbicides. However, it is difficult to produce bioherbicides that allow massive production at economical market range. This lack of progress may be accounted for by inadequate formulations and application technology. The former is thought to contribute to lower hindrance in bioherbicide development due to environmental factors. Fungal agents present as propagules that are applied with foliar need water for germination and penetration into weeds. This leaf wetness and its interaction with ambient temperatures influence the result of applying mycoherbicides. Sometimes, bioherbicides are evaluated by spraying till runoff and therefore are often over-estimated. Presence of dew in certain areas, therefore, alters the potential of mycoherbicides as they are very sensitive to desiccation and UV exposure.

10.2.2. Development of Mycoherbicides and Overcoming the Constraints

Generally, mycoherbicides are developed through screening of several fungal pathogens for their bioherbicidal potentials and then they are tested through biological and ecological assays for their feasibility as mycoherbicide candidates. Figure 10.1 shows a diagrammatic representation of a pragmatic approach that is needed for development of mycoherbicides. For mass production during fermentation, formulation of media highly influences the process. The shelf-life and environmental constraints for mycoherbicide need to be considered. Moreover, the media composition also affects the delivery and application of these mycoherbicidal agents. For example, if its too viscous, the fungal pathogen lose the ability to enter the target weed. There is an interrelation between mass-production, formulation, and application and hence when one of them is changed, the other two processes are influenced. Any modifications, refinements, and evaluations of either of these processes carried out during any stage of development may result in significant enhancements in performance of the mycoherbicide.

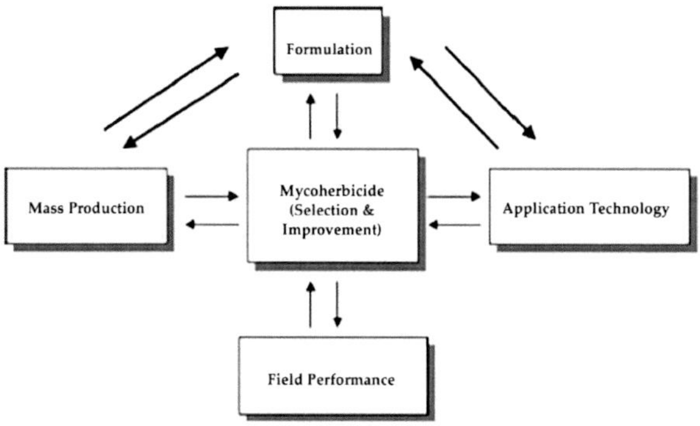

Figure 10.1: Framework for strategies employed for evaluating and developing mycoherbicides. Ward, G., (Source: *Hadar, Y., & Dosoretz, C. G. (2004). Fungal biotechnology in agriculture, food, and environmental applications.*).

10.2.3. Conclusion

When biological controls agents that are effective and active are identified, it is just the beginning of mycoherbicide product development. Selection of strain has to be continued to make sure that "nature's best" is used. Because there will always be limitations to using naturally occurring organisms, technologies like fermentation, formulation, and application will be necessary to determine the effectiveness of the agent as economically and commercially feasible mycoherbicidal product. To enhance its applicability in agriculture, it should be combined with chemical herbicides, traditional methods and employment of several pathogens to increase its flexibility and effectiveness in integrated weed management systems. It will also help decrease the chemical load imposed on the environment and on the crops. Keeping all of the above-discussed factors in mind, the individual mycoherbicide candidates are evaluated and realistic decisions regarding the presence of important characteristics needed for commercial application in agroecosystems. Conclusively, development of candidates for mycoherbicides is carried out by considering the following points;

1. Strains within a population of pathogens are assessed for variations and biodiversities on the basis of their epidemiological characteristics. These include their virulence, dispersal, infection efficiency and environmental adaptation;
2. Specific fungal agents are evaluated for large-scale production, with respect to fundamental elements related to their physical and

nutritionally requirements. This allows selection of ingredient substitutes that are economical during fermentation and for downstream processing methods;

3. Based on the mode of action and efficiencies of the candidate bioherbicides, relevant technologies for formulations are selected (e.g., liquid or solid matrices). To combat complex environmental challenges faced by fungal organisms, including UV exposure and temperature fluctuations, the formulation should consist of more than one ingredient;
4. To maximize efficiency of application, critical factors need to be considered in the application technology related to placement and penetration of the crop canopy on to the target. For example, factors like ability to overcome physical barriers like leaf cuticles and properties of leaf wetting, dispersal, and retention on the surface of leaves, optimized doses in the form of liquid droplets or solid granules, selection of appropriate technique for application like nozzles, position of angle or placement of application on soil (like side-banding). Along with formulation ingredients, the various parameters for application also need to be evaluated together;
5. The mycoherbicides are incorporated into crop production systems through various tools for week management. For example, they may be used synergically with chemical herbicides, or with other biological control agents or weed control options. This helps in optimizing the effectiveness of weed control. Within an integrated weed management system, the mycoherbicides need to be used sensibly to improve their value and applicability in order to control multiple weed species found in an agroecosystem.

10.3. BIOCONTROL AGENTS

Biocontrol agents (BCAs) are organisms that are used for controlling plant fungal diseases not directly or indirectly. These BCAs mat include bacterial or fungal species that are non-pathogenic and environmentally friendly. Biological control is something that occurs naturally, for example, the soil microbiome helps suppress wilt diseases caused by vascular wilt fungi (like *F. oxysporum*). When soil is enriched with organic matter in the form of compost, it adds nutrients into it and promotes the growth of microbe population. Some of these may act as disease suppressants. Therefore, biological control can be carried out by introducing foreign organisms or by altering the natural microbiome to support activity of already present organisms which suppress diseases. Fungal diseases are inhibited by microbes using three methods.

1. Direct: BCA grows directly around the pathogen. It suppresses its growth by mechanisms such as; production of anti-fungal metabolites or enzymes or interference by hyphae. For instance, several pathogenic fungi are biologically controlled by the fungus Trichoderma harzianum. This species produced hyphae and secretes degradative enzymes which interfere with fungal growth;
2. Indirect: In one mechanism, the pathogen is prevented from being in physical contact with the host by the help of BCA. It may also prevent pathogen growth by competing for nutrients within a particular area. For example, some non-virulent fungal strains like F. oxysporum present within the rhizosphere are able to out-compete certain virulent strains that cause vascular wilt diseases;
3. Induction of host defense: This mechanism is used by Pseudomonas spp. in various hosts to protect them from vascular wilt fungi. They do so by inducing SAR, one of the host defense mechanisms.

For a BCA to be effective, it must be developed and delivered to the host so that it may reach, in its active and viable form, the target niche within the ecosystem. It often happens that a BCA has high potential under experimental conditions but when employed under field conditions, it fails to be effective. It is a common practice to isolate BCAs from the same habitat where they are expected to be applied.

10.3.1. Control of Nematodes by Fungi

Nematodes are mostly saprotrophic roundworms that are found prevalently in many environments (aquatic, soil, and marine). These are parasites for both plants and animals resulting in damage to several crops and livestock. Nematicides are chemicals used for controlling nematodes that are plant pathogens but they often exhibit problems in the health and environment. Therefore, there is an increasing demand for eliminating nematodes using biocontrol agents on large-scale. Recently, there has been progress in studying nematophagous fungal species which may have potential to control spread of nematodes among plants. These fungi feed on nematodes through attacking, penetration, and digestions. They are able to do using several modes of actions;

1. Nematode-trapping fungi (Figure 10.2 A): They are able to trapvermiform nematodes within their hyphae by being sticky or through mechanical action. An example is of Arthrobotrys oligos;
2. Endoparasitic fungi (Figure 10.2 B): The fungal species are able to infect their host nematodes by production of spores that may be motile or non-motile. They are highly specific for their host for examples *Drechmeria coniospora*;
3. Toxin-producing fungi (Figure 10.2 C): These fungi, for example, Pleurotus ostreatus, immobilize their target nemaotes by using toxins and then penetrate into them.

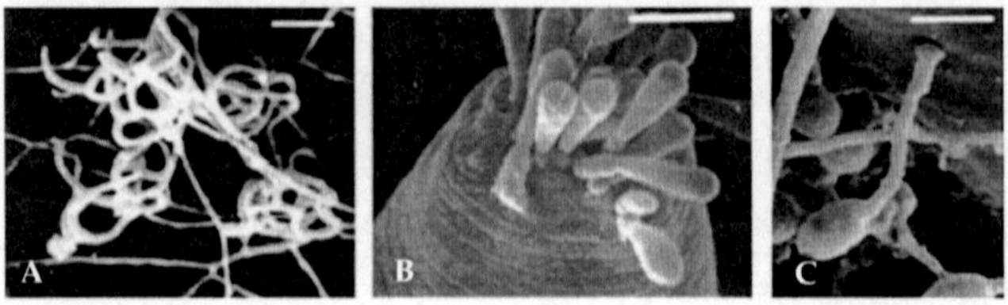

Figure 10.2: Scanning electron micrographs (SEM) of nematophagous fungi. A)Trap network produced by *Arthrobotrys oligospora*. B) *Drechmeria coniospora* producing adhesive conidia that is bounded to nematode head. C*) Pochonia rubescens* forming a protoappressorium on egg surface of *Heterodera schachtii*. Ward, G., (Source: *Hadar, Y., & Dosoretz, C. G. (2004). Fungal biotechnology in agriculture, food, and environmental applications*.).

10.3.2. Biofungicides

Many fungicides are commercially available which allow control of fungal diseases. Several of these contain variations of the same bioactive products. The fungicides can be used in the form of spray, paint, dust, soil treatment etc depending on the host and the disease type. Based on their method of penetration into plant tissue, their effects, and mode of actions, there are several types of biofunficides available.

Metabolites derived from microbes are of particular significance to control fungal diseases. This is because they have many versatile chemical structures, diverse biological activities, usually exhibit low cross-resistance towards commercial fungicides. In addition, they are biodegradable within few days or months lowering the chemical load on the environment.

Developments in techniques employed in the fields of chemistry, biology, and molecular biology have lead to facilitate screening for microbial metabolites that exhibit antifungal properties. The first biofungicide that was made commercially available was isolated from Blasticidin S which controls rice blast disease caused by Magnaporthe grisea. It functions by inhibition of growth of mycelia and germination of conidia in its target species. This major success led to researchers screening for more microbial fungicides and eventually it led to discovery and development of polyoxin, mildiomycin, kasugamycin, and validamycin which are produced by *Streptomyces cacaoi*, *Streptomyces rimofaciens*, *Streptomyces kasugaensis*, *Streptomyces hygroscopicus*, respectively. All of these biofungicides function by using various modes of actions, such as inhibition of proteins synthesis and production of fungal enzymes. They are classified into following groups:

1. **Protectant:** These fungicides affect viability and germination of pathogens without penetrating into the plant. They adhere to the surface of plants and resist weathering. They are effective as a preventative measure prior to fungal infection of plant;
2. **Systemic:** This category of biofungicides can act on surface of plant and are also able to translocate within the vascular system of plant to inhibit fungal growth. Systemic fungicides may also show translaminar movement within the leaves;
3. **Eradicative:** These biofungicides are used on already infected plants. When they come in contact with organism, they kill it or prevents its growth and reproduction.

10.3.3. Conclusion

Fungal species have a vast diversity in their life cycle, metabolism, and metabolic products. They are prevalently found in all the ecological niches. It is therefore expected that these organism have the ability to serve naturally as biocontrol agents. Due to this possible potential of fungi, they have been targeted in studies for finding their applications in various aspects of agriculture such as control of diseases caused by pests, insects, nematodes, fungi, bacteria, and various other pathogens. The fungi have proved to be useful systems for such applications to date. Due to fungal characteristics and with increasing awareness of significance of biocontrol agents in agriculture, it is expected that more and more fungal species will be identified and utilized by humans in this field.

REFERENCES

1. Aebersold R., Mann, M., (2003). Mass spectrometry-based proteomics. *Nature 422*, 198–207.
2. Arora, D. K., (2003). Fungal Biotechnology in Agricultural, Food, and Environmental Applications (Mycology).
3. Auerbach, D., Thaminy, S., Hottiger, M. O., Stagljar, I., (2002). The postgenomic era of interactive proteomics: Facts and perspectives. *Proteomics 2*, 611–623.
4. Baars, J., (2017). *Fungi as Food: Biology and Applications.*
5. Battilani, P., (2008). Food mycology – a multifaceted approach to fungi and food (Vol. 1).
6. Casselton, L., & Zolan, M., (2002). The art and design of genetic screens: filamentous fungi. [10.1038/nrg889]. *Nat Rev Genet, 3*(9), 683–697.
7. Causier, B., (2004). Studying the interactome with the yeast two-hybrid system and mass spectrometry (Vol. 23).
8. Cepononyte, S., Bruzyte, S., Melvydas, V., Serviene, E., (2008). Production of heterologous proteins using *S. Cerevisiae* expression system. *Biologija. 54*(3):178–182.
9. Curran, B., & Bugeja, V., (2017). *Biotechnological Exploitation of Heterologous Protein Production in Fungi: Biology and Applications.*

10. Davis, R. H., & Perkins, D. D., Neurospora: a model of model microbes. (2002). Nature, 3.
11. Davoli, T., Xu, W. A., Mengwasser, E. K., Sack, M. L., Yoon, J. C., Park, J. P. and Elledge, J. S., (2013). Cumulative Haploinsufficiency and Triplosensitivity Drive Aneuploidy Patterns and Shape the Cancer Genome. Cell.
12. Gellissen, G., Melber, K., A., Janowicz, Z., Dahlems, U., Weydemann, U., Piontek, M., Hollenberg, C., (1992). Heterologous protein production in yeast (Vol. 62).
13. Hartwell, L. H., (2002). Nobel Lecture: Yeast and Cancer. *Biosci Rep. 22*(3): 373–94. https://doi.org/10.1023/A:1020918107706.
14. Huang, B., Guo, J., Yi, B. et al., (2008). Heterologous production of secondary metabolites as pharmaceuticals in Saccharomyces cerevisiae. *Biotechnol Lett. 30*: 1121.
15. McKelvey, M. S., & Murphy, R. (2017). Biotechnological Use of Fungal Enzymes: Biology and Applications.
16. Moir, D., Stewart, S. E., Osmond, B. C., & Botstein, D., (1982). Cold-sensitive cell-division-cycle mutants of yeast: isolation, properties, and pseudoreversion studies. [Research Support, Non-U S Gov't. Research Support, US Gov't, P H S]. *Genetics, 100*(4), 547–563.
17. Murray, Patrick R., Rosenthal, Ken S., Pfaller, Michael A., (2009). Medical Microbiology/Philadelphia: Mosby/Elsevier.
18. O'Hanlon, R., (2017). *Fungi in the Environment: Biology and Applications*.
19. Paterson, R., (2005). Handbook of industrial mycology, vol. 22, Zhiqiang An, in: Mycology. Marcel Dekker, New York (2005), p. 763, ISBN: 0-8247-5655-X (Vol. 66).
20. Schierenberg, E., Miwa, J., & von Ehrenstein, G., (1980). Cell lineages and developmental defects of temperature-sensitive embryonic arrest mutants in Caenorhabditis elegans. Developmental Biology, 76(1), 141–159. doi: https://doi.org/10.1016/0012-1606(80)90368-1.
21. Spanu, P., (2005). Handbook of Fungal Biotechnology, second ed., D.Arora Marcel Dekker, Inc., New York and Basel (2004), 592, ISBN: 0-8247-4018-1 (Vol. 66).
22. Tedersoo, L., Nilsson, R. H., Abarenkov, K., Jairus, T., Sadam, A., Saar, I., Kõljalg, U. (2010). 454 Pyrosequencing and Sanger sequencing of

tropical mycorrhizal fungi provide similar results but reveal substantial methodological biases. New Phytologist, 188(1), 291–301. doi: 10.1111/j.1469-8137.2010.03373.x.

23. Ward, G., Hadar, Y., & Dosoretz, C. G. (2004). Fungal biotechnology in agriculture, food, and environmental applications.

INDEX

C

D

G

H

I

K

L

M

N

O

P

Q

R

S

T

U

V

W

Z

Printed and bound by PG in the USA